# RED PLANET

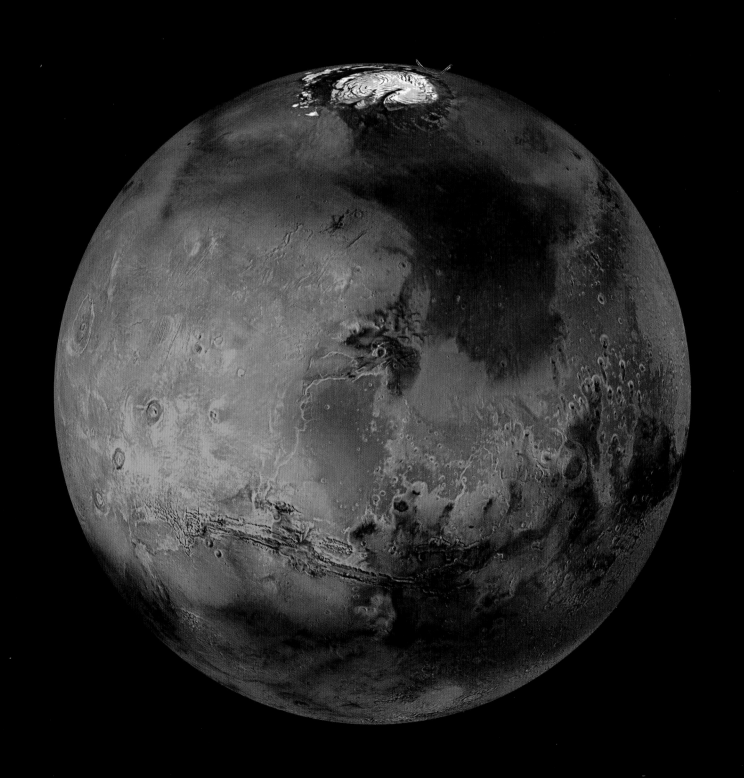

STERLING
New York

An Imprint of Sterling Publishing
387 Park Avenue South
New York, NY 10016

ISBN 978-1-4549-1780-9

This book was designed, conceived and produced by
Quintet Publishing Ltd
4th Floor, Sheridan House
112/116A Western Road
Hove, East Sussex
BN3 1DD

Design and editorial content: Tall Tree Ltd
Art Director: Michael Charles
Editorial Director: Emma Bastow
Editorial Assistant: Ella Lines
Publisher: Mark Searle

Distributed in Canada by Sterling Publishing
c/o Canadian Manda Group, 165 Dufferin Street
Toronto, Ontario, Canada M6K 3H6

For information about custom editions, special sales, and premium and corporate purchases, please contact Sterling Special Sales at 800-805-5489 or specialsales@sterlingpublishing.com.

Manufactured in China

2  4  6  8  10  9  7  5  3  1

www.sterlingpublishing.com

# RED
# PLANET

## GILES SPARROW

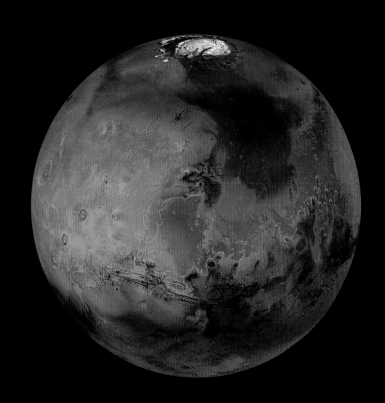

STERLING
New York

# CONTENTS

# INTRODUCTION

Mars has exerted a powerful fascination on the human imagination since before our ancestors developed written language. Constantly changing in position and brightness, it is the least predictable of all the "wandering stars" known as planets. At its brightest it can outshine almost every other object in the night sky, while at its faintest it is outshone by a dozen or more of the brighter stars. Yet one thing renders Mars almost unmistakable whatever its brightness—its color.

## The Color of Blood

No one knows who first referred to Mars as the Red Planet, but the name certainly stuck. Its distinctive, ruddy color makes it perhaps the most famous planet of all, yet this color—the color of blood, rage, and war—has always given Mars balefully superstitious associations that continue almost to the present day.

The first people ever to study Mars were probably the Sumerians, who developed the first form of writing, cuneiform, as long ago as 3500 BCE and used a sexagesimal number system, which is based on 60 place values, in much the same way that our decimal system is based on 10. This number system made it easier for them to express very large numbers as well as very small fractions, which was essential when they came to try to catalog the movements of the stars and planets. Sumerian astronomy evolved as "astrolatry" where the stars and planets are worshipped as gods or in association with certain gods. This was a tradition that would be carried down throughout the millennia.

The earliest written records and depictions of Mars come from the astronomers of ancient Egypt and Babylon. Babylonian astronomers began using mathematics to follow and predict the way the stars and planets moved across the sky. They based their calculations on empirical knowledge from centuries of recorded observation, and there is evidence that they used Sumerian research as the names of heavenly bodies carved on surviving Babylonian stone tablets often appear in Sumerian.

## Egyptian Stargazers

As early as the 15th century BCE, the tomb of Senenmut, court astronomer to the famous Queen Hatshepsut, illustrates a conjunction (coming-together) of planets in the sky that occurred in 1534 BCE, with the position of Mars marked by an empty boat.

The Egyptians undoubtedly tracked the motion of the planets long before this, and were certainly well aware of the Red Planet's unusual pattern of movement.

RIGHT An Egyptian statue showing Senenmut, the court astronomer to Queen Hatshepsut, with the Queen's daughter Neferure on his lap. Based on the decorated ceiling of his tomb, Senenmut is the earliest individual known to have taken an interest in Mars.

ABOVE Babylonian mythology imagined fearsome gods with a variety of attributes such as wings and the bodies of bulls. Nergal, the Babylonian god of war and plague associated with Mars, was often depicted as a lion or bearing a lion-headed mace.

## Planet in Retrograde

While Jupiter and Saturn track fairly steadily westward against the fixed constellations of the zodiac, and Mercury and Venus execute narrow and wide circles around the Sun, Mars follows a generally westward track, but very obviously reverses its motion for several months every couple of years.

Today, we understand that these "retrograde loops" result from the relative motion of Earth and Mars in their orbits— indeed, the struggle to understand the planet's frustrating motions was ultimately to unlock a revolution in astronomy. To ancient stargazers, however, the unpredictability and volatility of Mars undoubtedly added to its fascination.

## The War God

The first hints of warlike associations, and the ancestry of the planet's modern name, are to be found in ancient Mesopotamia almost a thousand years after Senenmut. Here, Babylonian astronomers developed the first detailed mathematical models of planetary motion, driven by a fascination with their own particular version of astrology.

Anticipating the behavior of Mars and the other planets became a matter of considerable interest. The complex

unpredictable Red Planet was particularly important because the Babylonians associated it with the deity Nergal—Mesopotamian god of the Underworld, and a fearsome harbinger of fire, destruction, and war.

## A Lasting Legacy

The flowering of the Babylonian Empire was short-lived, but it left a lasting scientific legacy. The 360 degrees into which we divide a circle are a Babylonian invention, as are the 24 hours in a day, as well as the counting system we use to divide hours into minutes and minutes into seconds.

These systems came down to us through the ancient Greeks, the founders of Western civilization but also the inheritors of much earlier Mesopotamian culture—and the Greeks also seem to have inherited the association of the Red Planet with a warrior god.

RIGHT The baleful face of Mars, revealed in a mosaic of about a thousand Viking orbiter images. The Red Planet's brilliance, distinctive color, and unpredictable motion across the sky led astrologers throughout history to attribute it with unusual powers. Its associations with war have persisted into our more scientific age, in works of art ranging from Gustav Holst's Planets Suite to H. G. Wells' *War of the Worlds*.

## The God of War

From at least the 5th century BCE, the Greeks saw the planet as an aspect of Ares, god of war. Although often regarded as synonymous with the Roman god Mars, Ares was considered far less noble in the Greek pantheon—the nobler aspects of war were attributed to his sister Athena, while Ares embodied violence and brutality, and was frequently treated with scorn in Greek mythological tales.

Mars, in contrast, was considered second only to Jupiter himself among the gods of Rome, and regarded as a protector and guardian of the city and its famously militaristic people. The planet's symbol (also used as a universal icon for maleness) is a stylized representation of Mars' shield and arrow.

Other cultures have independently associated the planet with war and bloodshed. In Hindu cultures it is the warrior god Mangala, and in China it is regarded as the "fire star," again seen as a portent of war and destruction.

The Germanic and Scandinavian peoples of northern Europe may have made similar associations with their war-god Tiw or Tyr. The second day of the week is named Tuesday in his honor among Anglo-Saxon cultures, and it is surely no coincidence that the Romans dedicated this same day of the week to Mars, still recalled in the French "Mardi."

Throughout the European Middle Ages and the heyday of Islamic astronomy, Mars was seen as a powerful astrological force, but with the European Renaissance, things slowly began to change. Advances in theory and new observing tools brought the real Mars within reach for the first time.

ABOVE This spectacular sculpture from the Forum of Nerva in Rome depicts Mars as a majestic warrior god, a symbol of masculinity, and a fitting deity for a people who, at their peak, conquered much of the known world.

Ancient Greek and Islamic astronomers had attempted to measure basics such as its diameter and distance, some even recognizing that, in principle, Mars was a body not too different in size from Earth, but it was the arrival of the telescope that began to reveal the Red Planet as a physical world in its own right.

## Mars as We Know It

It is the physical Mars that is the subject of this book—the fourth planet of our solar system and the one most similar to Earth. *The Red Planet* is both an introduction to the wonders of Mars, and a history of their discovery.

In five chapters, we travel from the world of pre-telescopic astronomy, where the greatest concern was the way Mars moved in the sky, to a future of manned landings, and perhaps even large-scale Martian colonies reshaping the planet in our own image.

Chapter 1, "The Discovery of Mars" describes in detail the pivotal struggle to understand the movement of Mars, and its implications for the development of astronomy as a whole. It also looks at the earliest telescopic discoveries about the planet, and the various efforts to map its surface by Earth-based astronomers in the 19th and 20th centuries.

Chapter 2, "First Reconnaissance" chronicles the first space missions to Mars. From basic robot probes in the early 1960s, which were barely capable of flying past the planet while still sending back signals, to the sophisticated Viking orbiters and landers of the mid-1970s.

Along the way, we also look at some of the Red Planet's landmark features—the giant volcanoes and vast canyons that are an integral part of modern Martian mythology.

Chapter 3, "Mapping the Red Planet" looks in detail at the second generation of Mars orbiters and the discoveries they have made. Following a long hiatus, this began with Mars Global Surveyor's successful entry into Martian orbit in September 1997, and has continued to the present day. Along the way, advanced cameras and other ingenious scientific instruments have transformed our understanding of Mars past and present, revealing ever-increasing similarities to Earth.

Chapter 4, "Curiosity and its Kin" examines the surface landers and rovers that are supplementing the view from orbiters with close-up studies of the atmosphere, soil, and geology. Mars Pathfinder, the pioneer of this approach, arrived weeks before Global Surveyor in 1997, and today its descendants, the

**ABOVE** The space age has transformed our view of Mars. Today we can envisage the planet's landscape in amazing detail. Perspective views such as this one of the Acidalia Planitia region, captured by the European Space Agency's Mars Express spacecraft, bring us closer to the Red Planet than ever before.

long-lived Opportunity and ambitious Curiosity, are still sending back a bounty of data.

Chapter 5, "Onward to Mars" tells the story of the Red Planet's future, from the hottest current prospects for the discovery of life on Mars, through recently arrived and planned orbiters and landers, to ambitious plans for manned Martian landings and the establishment of a permanent colony.

## Spotting the Red Planet

Today, we know more about Mars than ever before, but this only serves to make the sight of the Red Planet all the more fascinating for amateur stargazers. Except around conjunction, when it passes behind the Sun as seen from Earth, Mars is always visible somewhere in the sky as a bright red "star."

Moving generally westward against the zodiac constellations, it completes an orbit of the Sun every 1.88 Earth years. It appears first in the evening sky after sunset, then gradually draws away from the Sun until around the time of opposition, when Mars is brightest in the sky and closest to Earth, visible all night, rising in the east as the Sun sets in the west.

This is the time when Mars describes a large retrograde loop in the sky, moving eastward for several weeks before resuming its westward course. The Red Planet then starts to disappear from evening skies, rising later and later until, as it approaches solar conjunction once again, it rises just before the Sun in the pre-dawn eastern sky. Thanks to the complex interplay of Earth and Martian orbits, the entire cycle takes more than one Martian orbit to complete.

Each conjunction or opposition occurs an average of two years and 49 days after the last—taking place, therefore, some way to the west of the previous one in the zodiac.

Mars is always at its brightest around opposition and its faintest near conjunction, but its brightness varies considerably from one cycle to the next, particularly around opposition, where the closest separation of Earth and Mars can vary considerably thanks to the elliptical Martian orbit.

The table opposite shows the best times to see Mars over the next decade, along with its peak brightness at oppositions. This is measured in terms of magnitude, a scale in which lower numbers indicate brighter objects. For comparison, the brightest star in the sky, Sirius, has a magnitude of -1.4, while the faintest naked-eye stars have magnitudes around 6.0.

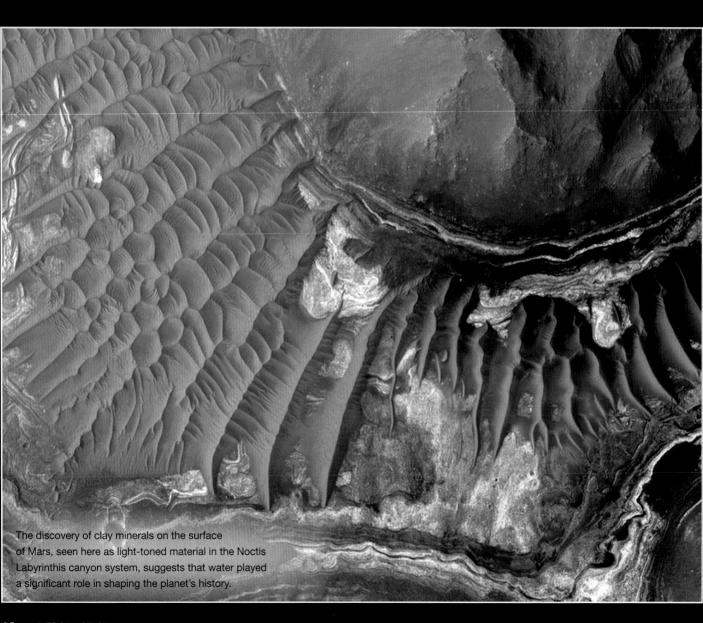

The discovery of clay minerals on the surface of Mars, seen here as light-toned material in the Noctis Labyrinthis canyon system, suggests that water played a significant role in shaping the planet's history.

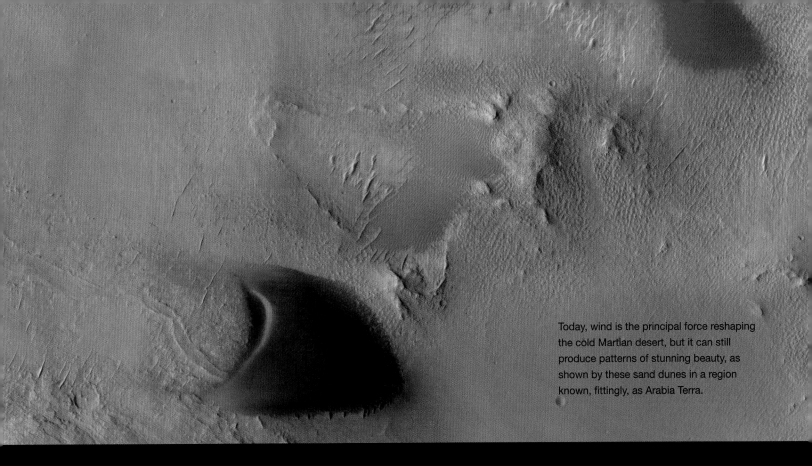

Today, wind is the principal force reshaping the cold Martian desert, but it can still produce patterns of stunning beauty, as shown by these sand dunes in a region known, fittingly, as Arabia Terra.

| Date | Visibility | Constellation |
| --- | --- | --- |
| Aug 2015 – Apr 2016 | Evening -> All night | Gemini -> Libra |
| May 2016 | Opposition (mag -2.0) | Scorpius |
| Jun 2016 – May 2017 | All night -> Morning | Sagittarius -> Gemini |
| Jun – Aug 2017 | Conjunction (not visible) | Cancer |
| Sep 2017 – Jun 2018 | Evening -> All night | Leo -> Sagittarius |
| July 2018 | Opposition (mag -2.8) | Capricornus |
| Aug 2018 – Jul 2019 | All night -> Morning | Aquarius -> Cancer |
| Aug – Oct 2019 | Conjunction (not visible) | Leo |
| Nov 2019 – Sep 2020 | Evening -> All night | Virgo -> Aquarius |
| Oct 2020 | Opposition (mag -2.6) | Pisces |
| Nov 2020 – Aug 2021 | All night -> Morning | Aries -> Leo |
| Sept – Nov 2021 | Conjunction (not visible) | Virgo |
| Dec 2021 – Nov 2022 | Evening -> All night | Libra -> Aries |
| Dec 2022 | Opposition (mag -1.8) | Taurus |
| Jan 2023 – Sep 2023 | All night -> Morning | Gemini -> Virgo |
| Oct – Dec 2023 | Conjunction (not visible) | Libra |
| Jan – Dec 2024 | Evening -> All night | Scorpius -> Taurus |
| Jan 2025 | Opposition (mag -1.4) | Gemini |
| Feb – Nov 2025 | Evening -> All night | Cancer -> Scorpius |
| Dec 2025 – Feb 2026 | Conjunction (not visible) | Sagittarius |
| March 2026 – Jan 2027 | Evening -> All night | Capricornus -> Cancer |
| February 2027 | Opposition (mag -1.2) | Leo |

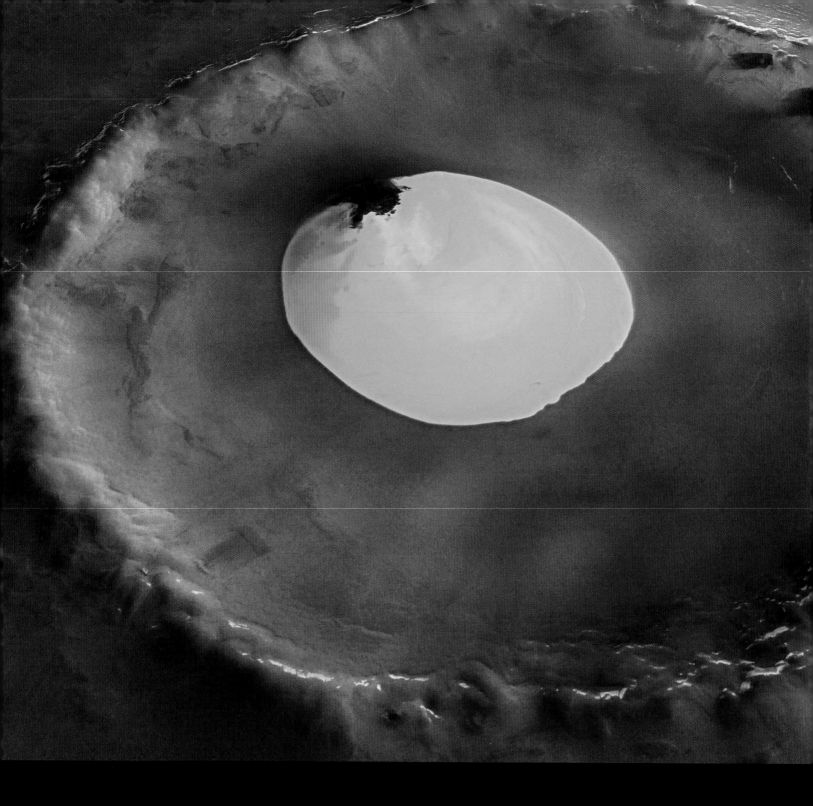

## Full of Surprises

Just as the brightness of Mars varies through the cycle of oppositions and conjunctions, so does its apparent size in Earth's skies. At close oppositions it can reach an "angular diameter" of about 25 seconds of arc (roughly 1/60th the diameter of a full Moon), while at other times it can dwindle to less than one-sixth that size.

Despite its proximity and brightness, Mars sadly appears as little more than a bright red star when viewed through binoculars. A decent telescope with high magnification is needed to transform it into a ruddy disc and start to reveal some of its surface features.

It might seem that, in an age where orbiting space probes have mapped the entire surface of Mars, Earth-based astronomers can make little more contribution to our scientific knowledge of the Red Planet.

In fact, Mars still has the potential to surprise. Even the most sophisticated probes can sometimes miss important details, while stay-at-home astronomers can get a unique overview of

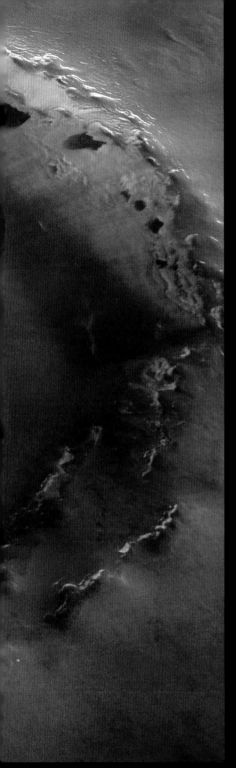

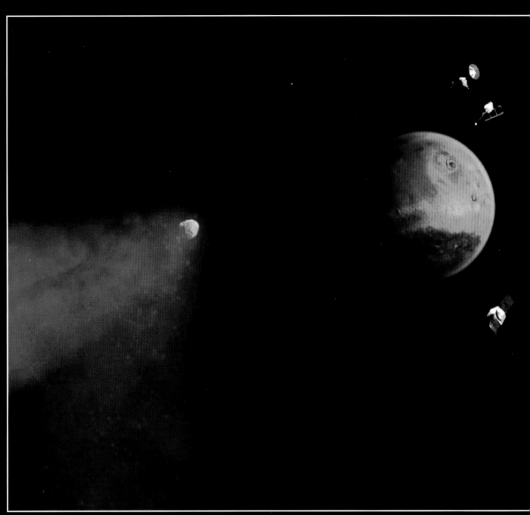

ABOVE A close encounter: In late 2014 NASA maneuvered its Mars orbiters into safe positions on the far side of Mars during the close approach of Comet Siding Spring. The small comet came within 87,500 miles (140,000 km) of Mars (about one-third of the distance between Earth and Moon), enriching the atmosphere with a variety of elements from its dust and ice, but otherwise leaving it, and the orbiting spacecraft, unharmed.

LEFT This famous image of an ice lake within a crater near the Martian north pole offers stunning confirmation that the Red Planet is not the dry, barren world we once thought.

the entire planet at any time. The first hints of methane in the Martian atmosphere, one of the most exciting discoveries of our time, were found in this way, so even backyard observers can make surprising discoveries.

Professional astronomers are still puzzling over the significance of an enormous bright plume spotted over the planet's southern hemisphere in 2012. Could it be a strange form of Martian aurora australis, or a sudden eruption of carbon dioxide? The phenomenon would not have been noticed at all without the efforts of eagle-eyed enthusiasts and

it was, after all, just such eruptions that heralded an alien invasion in H. G. Wells' classic 1897 story *War of the Worlds*.

The pages of this book tell the story of our long obsession with Mars, while the images capture some of its astounding alien beauty. But there can be nothing better than seeing it for yourself, even with the unaided eye.

Next time you are under clear, dark skies, take a moment to track Mars down, and relish the sight of a world that has obsessed humanity for millennia—the Red Planet.

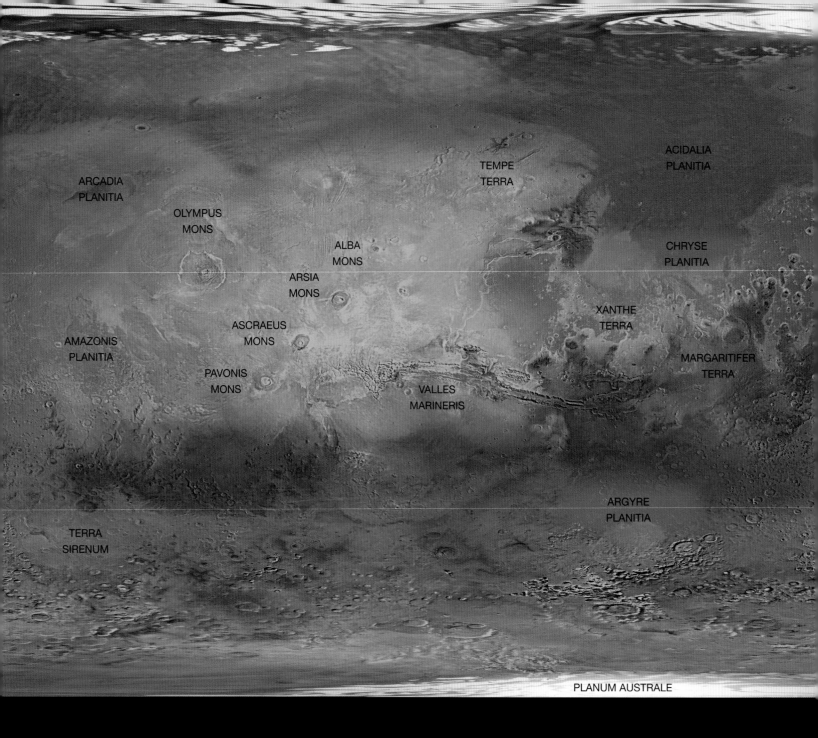

ARCADIA
PLANITIA

OLYMPUS
MONS

ALBA
MONS

ARSIA
MONS

ASCRAEUS
MONS

AMAZONIS
PLANITIA

PAVONIS
MONS

VALLES
MARINERIS

TERRA
SIRENUM

TEMPE
TERRA

ACIDALIA
PLANITIA

CHRYSE
PLANITIA

XANTHE
TERRA

MARGARITIFER
TERRA

ARGYRE
PLANITIA

PLANUM AUSTRALE

# Mapping Mars

The map displayed here is our most comprehensive global
look at the Red Planet—a huge patchwork combining the best
detail from around 4,600 images captured by NASA's Viking
Orbiter spacecraft. Compiled by the United States Geological
Survey (USGS) and further tweaked by scientists at NASA's
AMES Research Center, the result is known as the Mars Digital
Image Model (MDIM) 2.1.

In the original electronic version, each square degree of the
planet's surface is represented by a square 256 pixels wide
and tall, corresponding to a scale of 758 feet (231 meters) per
pixel at the equator, which improves at higher latitudes. The
entire map is printed here in a form known as cylindrical

projection. All map projections introduce distortions, but a
cylindrical projection is one of the easiest types of map to
interpret, and confines the worst of the distortion to high
latitudes. The color elements of the map, meanwhile, attempt
to replicate the true appearance of Mars by combining
monochrome information about surface brightness captured
through several separate filters. Variations in the angle of
sunlight between different areas of the image have been
removed using computer processing techniques.

Despite the success of more recent Martian orbiters equipped
with far higher-resolution cameras, the MDIM remains a
benchmark because it is the result of a truly global survey—
other space probes have imaged only small areas of the planet

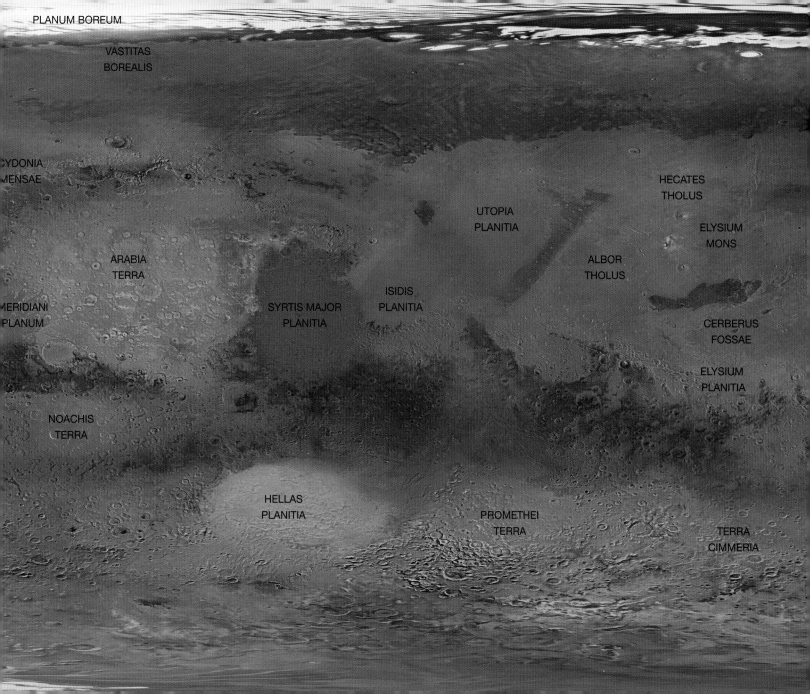

PLANUM BOREUM

VASTITAS
BOREALIS

CYDONIA
MENSAE

HECATES
THOLUS

UTOPIA
PLANITIA

ELYSIUM
MONS

ARABIA
TERRA

ALBOR
THOLUS

ISIDIS
PLANITIA

MERIDIANI
PLANUM

SYRTIS MAJOR
PLANITIA

CERBERUS
FOSSAE

ELYSIUM
PLANITIA

NOACHIS
TERRA

HELLAS
PLANITIA

PROMETHEI
TERRA

TERRA
CIMMERIA

# THE DISCOVERY OF MARS

For most of human history, our only means of studying Mars (and the heavens in general) was with the unaided eye. But even with these limitations, early stargazers learned a surprising amount about the Red Planet—and ultimately it was observations of Mars that led to a revolution in our understanding of the Universe as a whole. Then, with the invention of the telescope in the early 17th century, new discoveries began to turn Mars from a light in the sky to a planet in its own right—a world with geographical features, weather systems, seasons, and two satellites. Perhaps, some thought, our neighboring planet might even harbor intelligent alien life. Today, while space probes have transformed our view of Mars beyond all recognition, observations from Earth still have an important role to play in our understanding.

People have stared at the stars and tried to make sense of them since long before the dawn of written language. The first astronomers told stories to make sense of the heavens, and would have realized early on that there were five bright "stars" that didn't behave like all the rest. While the vast majority of lights in the sky remained fixed in place from night to night, and from year to year, these five celestial wanderers changed their brightness and location. We now know them as planets, from the Greek for "wanderers," and understand them to be relatively nearby neighbors in space—the brightest and most obvious members of our solar system, each in its own orbit around our nearest star, the Sun.

For a long time, however, this simple truth was far from obvious, and the discovery of the structure of our solar system was led by observations of Mars in particular. The Red Planet was prominent because of its color, but also because of its unpredictability—it went through complex changes in both brightness and motion through the sky. Sometimes it could outshine all the other "stars" except for the brilliant Venus, while at other times it dwindled in importance beneath a dozen or more stars and planets.

## Looping Through the Sky

Mars' path across the heavens was even more puzzling. Along with the Sun and other planets, it remained confined within a narrow band of the sky, only ever appearing against the star patterns or constellations of the zodiac. The Sun tracked steadily eastward against this band throughout the year, following a line known as the ecliptic, but the planets behaved differently. Mercury and Venus only ever appeared close to the Sun, inscribing loops to the west or east and appearing as morning or evening stars before reversing their

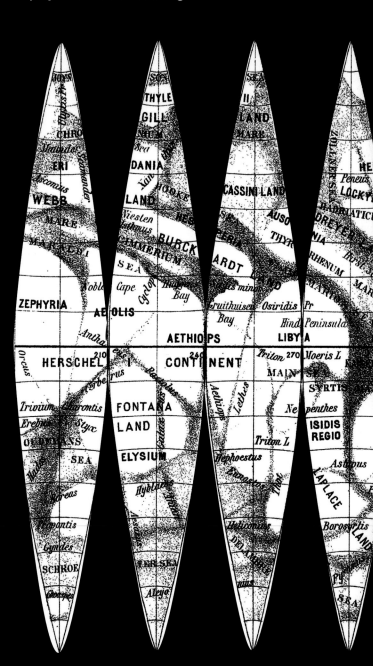

RIGHT This 1892 map of Mars by Belgian astronomer Louis Niesten is designed to be folded and joined into a globe. The names used are largely obsolete—today's names are based on the work of Greek astronomer Eugène M. Antoniadi in the 1920s.

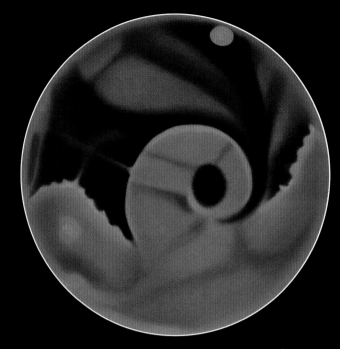

motion and disappearing back into the twilight. Jupiter and Saturn tracked more or less steadily across the sky, following westward paths that progressively took them through the entire band of constellations over periods of about 12 and 30 years, respectively.

Mars, on the other hand, was noticeably different. While it generally followed the same westward track as Jupiter and Saturn, it moved considerably faster and would periodically reverse its direction, inscribing large "retrograde" loops in the sky. As instruments for measuring positions in the sky improved with the invention of devices such as astrolabes and quadrants, it became clear that Jupiter and Saturn also made retrograde loops, albeit much smaller and more subtle ones.

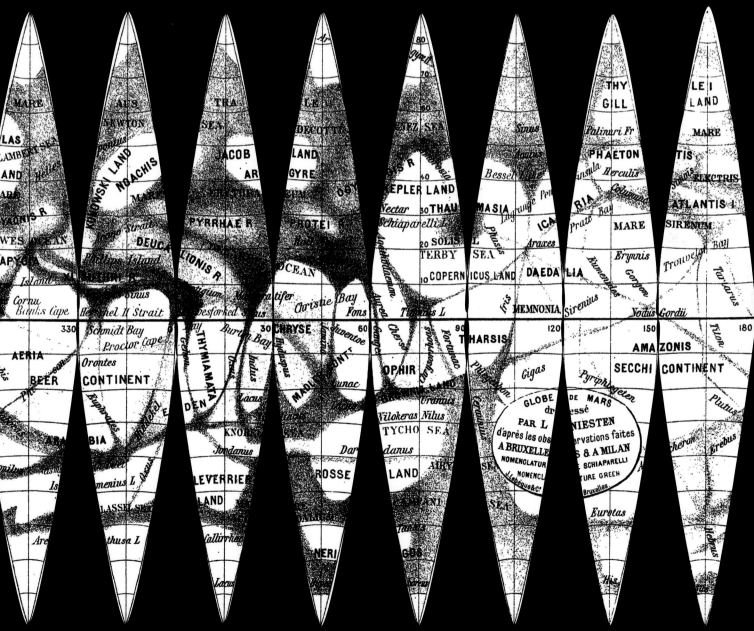

## The Importance of Mars

Accurate measurements simply added to the complexities of Martian movement. Unlike the loops of Jupiter and Saturn, those of Mars could be of markedly different size and duration.

Throughout this early period, and right up until the 17th century, the studies of astronomy (the science of the stars and planets) and astrology (the belief that the motions of celestial objects influence events on Earth) were closely linked.

Kings and emperors sponsored observatories as well as individual astronomers and astrologers at least in part because they believed that the positions of the stars and planets would influence the outcome of decisions they had to make. An astronomer's observations or the way an astrologer interpreted the movement of heavenly bodies might even foretell the fate of a king. Medieval stargazers were, therefore, highly influential and bore huge responsibility.

This made Mars an enormously important planet. Its strange, unpredictable motion created a source of serious concern for those who believed that the movement of the stars and planets had a direct effect on terrestrial events.

Today, of course, we understand that Mars is the fourth of eight major planets in our solar system. Orbiting between Earth and Jupiter, it is the outermost and second-smallest of the so-called "terrestrial" or rocky planets, with a diameter of 4,245 miles (6,792 km) compared to Earth's 7,923 miles (12,756 km). Directly beyond Mars orbits the asteroid belt, a broad ring of small, rocky worlds dominated by the dwarf planet Ceres, a mere 594 miles (950 km) across.

The asteroids are thought to be debris left over from the formation of the solar system, shepherded into their present paths by the gravitational influence of Mars on the inner edge, and the far greater effects of Jupiter, the first and largest of the solar system's giant planets, orbiting beyond them.

## Gas Cloud Collisions

Along with the rest of the solar system, Mars is thought to have formed around 4.6 billion years ago from a collapsing

cloud of interstellar gas and dust. This "solar nebula" was made up of the same kind of material seen today in famous star formation regions such as the Orion and Carina nebulae.

Our solar system formed from a clump of matter that became separated from a larger cloud and began to collapse under its own gravity. As mass became concentrated toward the center, it began to spin more rapidly (rather like a pirouetting ballerina pulling her arms in toward her body), while collisions between gas clouds moving in different directions gradually flattened the solar nebula out into a disk with a bulging center.

## Nuclear Fusion in Space

As the very heart of the nebula pulled more and more material inward through gravity, it accumulated more and more mass, growing steadily denser and hotter.

Eventually, conditions became so extreme that the nuclei of hydrogen atoms (the lightest and most common element in the Universe) began to collide and stick together, releasing energy through a process known as nuclear fusion. The center

of the nebula had become a newborn star. As this infant Sun began to pump heat and light out into the solar system, it had dramatic effects on the material in the surrounding ring.

## Ice and Dust

Broadly speaking the debris that made up the ring consisted of any remaining gas clouds, along with chemical compounds that had relatively low melting points. These can be termed "ices," and include water ice. Material with a higher melting point is generally known as "dust."

Close to the Sun, solar radiation melted nearby ice into vapor, leaving only dust particles orbiting in a cloud of gas that was steadily driven outward by the pressure of radiation and the "solar wind" (streams of particles blowing out from the Sun's hot surface). Farther out, beyond the "snow line" of the early solar system, vast amounts of ice persisted amid the gas.

BELOW This artist's rendition shows the major objects of our solar system (though not to scale). Mars' position at the edge of the inner solar system, and relatively small size, are clear.

## Forming the Planets

According to the latest thinking, the planets formed not by the steady collision and conglomeration of small particles to form increasingly large ones, but by the relatively sudden coming-together of huge drifts of pebble-sized particles. These drifts were created rather like the peloton of a bicycle race as pebbles streamed behind each other, swept into place by the headwinds from the gas through which they were moving.

When the drift grew large enough to become gravitationally unstable, it collapsed to form objects over 1,000 miles (1,600 km) across. That process is believed to have taken just a few years. These "planetesimals" then continued to grow rapidly as their gravity swept up the remaining pebbles from their surroundings.

Computer models suggest this so-called "pebble accretion" process may have produced planets the size of Mars within a couple of million years of the Sun forming. In the ice-rich outer solar system, much larger, giant-planet cores would have formed.

Larger rocky planets like Earth may have required a further stage in growth by collisions between planetesimals—explaining why geological studies suggest our own planet was born about 100 million years after Mars.

## Evolving Orbit

Today, Mars' orbit is one of the least uniform among the planets. Its distance from the Sun varies between 129.2 million miles (206.7 million km) at perihelion, which is its closest point to the Sun, and 155.7 million miles (249.2 million km) at its most distant point, or aphelion.

According to most theories of planet formation, all planets should have started their lives in more or less circular orbits, so the eccentricity of Mars' current orbit is something of a mystery. What we do know is that, under the gravitational influence of other planets, Mars' orbit actually flexes its shape between an elongated, elliptical path and a far more regular, close-to-circular route in a 124,000-year cycle.

## The Martian Year

Mars orbits the Sun once every 687 Earth days. This makes a Martian year equivalent to 1.88 Earth years. It is this slower orbit that produces the planet's distinctive motion in Earth's skies. Because Mars is farther from the Sun, it not only has farther to travel to complete an orbit, but it also moves more

2010

61.9 million miles
(99 million km)

2012

2014

2016

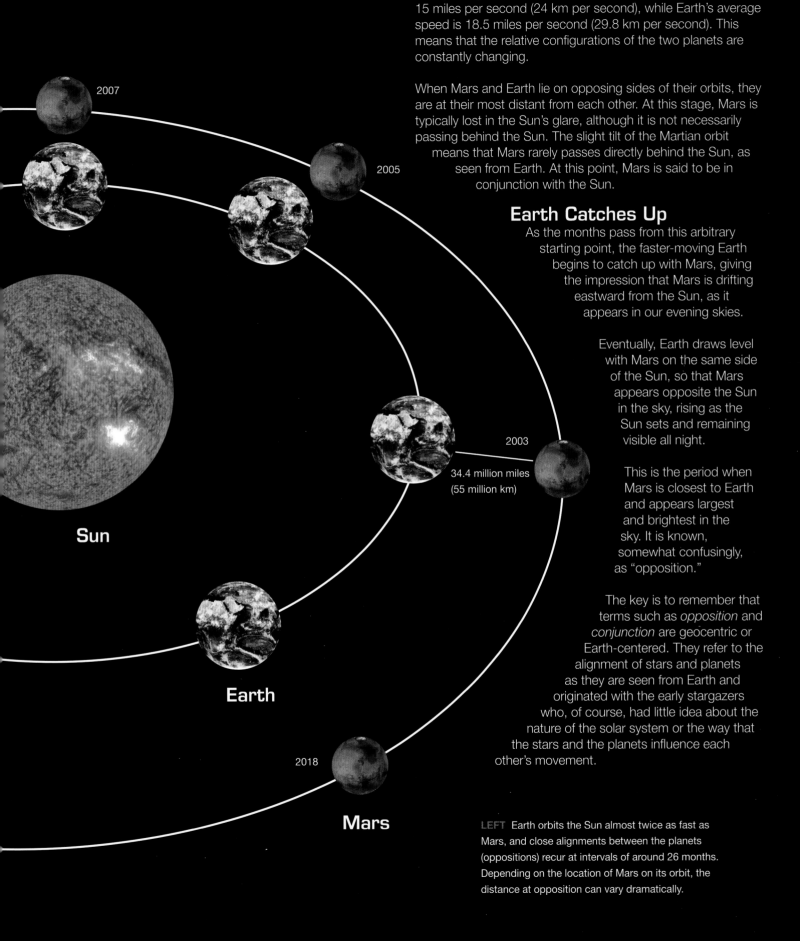

slowly along its path. On average, it moves through space at 15 miles per second (24 km per second), while Earth's average speed is 18.5 miles per second (29.8 km per second). This means that the relative configurations of the two planets are constantly changing.

When Mars and Earth lie on opposing sides of their orbits, they are at their most distant from each other. At this stage, Mars is typically lost in the Sun's glare, although it is not necessarily passing behind the Sun. The slight tilt of the Martian orbit means that Mars rarely passes directly behind the Sun, as seen from Earth. At this point, Mars is said to be in conjunction with the Sun.

## Earth Catches Up

As the months pass from this arbitrary starting point, the faster-moving Earth begins to catch up with Mars, giving the impression that Mars is drifting eastward from the Sun, as it appears in our evening skies.

Eventually, Earth draws level with Mars on the same side of the Sun, so that Mars appears opposite the Sun in the sky, rising as the Sun sets and remaining visible all night.

This is the period when Mars is closest to Earth and appears largest and brightest in the sky. It is known, somewhat confusingly, as "opposition."

The key is to remember that terms such as *opposition* and *conjunction* are geocentric or Earth-centered. They refer to the alignment of stars and planets as they are seen from Earth and originated with the early stargazers who, of course, had little idea about the nature of the solar system or the way that the stars and the planets influence each other's movement.

2007

2005

2003

34.4 million miles
(55 million km)

Sun

Earth

2018

Mars

**LEFT** Earth orbits the Sun almost twice as fast as Mars, and close alignments between the planets (oppositions) recur at intervals of around 26 months. Depending on the location of Mars on its orbit, the distance at opposition can vary dramatically.

It was these complications that made Mars such a problem for early European astronomers. Up until the 16th century, most stargazers agreed with the general principles of a model that had been handed down since ancient Greek times—the Earth-centered or geocentric Universe.

Developed to its highest level by the Greek-Egyptian polymath, Ptolemy of Alexandria, around 150 CE, this model placed the Earth at the center of everything. The model explained how the Sun, Moon, and planets orbited around the Earth on crystalline spheres that moved at different speeds, while an outer shell of stars surrounded the entire cosmos.

In order to explain phenomena such as the retrograde loops of Mars and the tendency of Mercury and Venus to stay close to the Sun, Ptolemy introduced a complex system of sub-orbits known as epicycles. As observation techniques improved across both the Islamic and Medieval European worlds, it became clear that Ptolemy's model was consistently failing in its predictions. By this time, however, the Earth-centered Universe had become fundamental to both academic and religious teaching.

In the early 1500s, Polish priest Nicolaus Copernicus began to develop an alternative theory—a "heliocentric" Universe in which Earth was just one of several planets orbiting the Sun.

ABOVE This portrait of Johannes Kepler was painted in 1610 by an unknown artist. The German mathematician and astronomer had many other interests, including music and astrology.

## The Lazy Loop

It's around the time of opposition that Mars makes its retrograde motion in the sky. As the relatively speedy Earth overtakes it on the inside lane, from our point of view Mars appears to move backward, describing a lazy loop before eventually resuming its eastward course. Finally, it begins to approach the Sun again, disappearing from the evening skies and becoming a morning object before it eventually vanishes into the pre-dawn glare around the Sun, completing the cycle.

In total, Mars takes an average of 780 days (its "synodic period") to complete an entire circuit of Earth's skies and return to the same position relative to the Sun, so this period (roughly two years and two months) is the average interval between bright oppositions.

The exact details of Mars' synodic cycle are complicated by its eccentric orbit. While Earth's path around the Sun is more or less circular, and our planet maintains a fairly steady speed in its orbit, Mars' more elongated ellipse makes it speed up when closer to the Sun, and slow down when further away. This can cause oppositions to repeat sooner than would otherwise be expected, or it can delay them, and can also produce radical differences in the apparent brightness of Mars from one opposition to the next. When an opposition coincides with Martian perihelion, for example, the distance between Earth and Mars can be as little as 34.4 million miles

ASTRONOMIA NOVA
ΑΙΤΙΟΛΟΓΗΤΟΣ,
SEV
PHYSICA COELESTIS,
tradita commentariis
DE MOTIBVS STELLÆ
MARTIS,
Ex obſervationibus G. V.
TYCHONIS BRAHE:

Juſſu & ſumptibus
RVDOLPHI II.
ROMANORVM
IMPERATORIS &c:

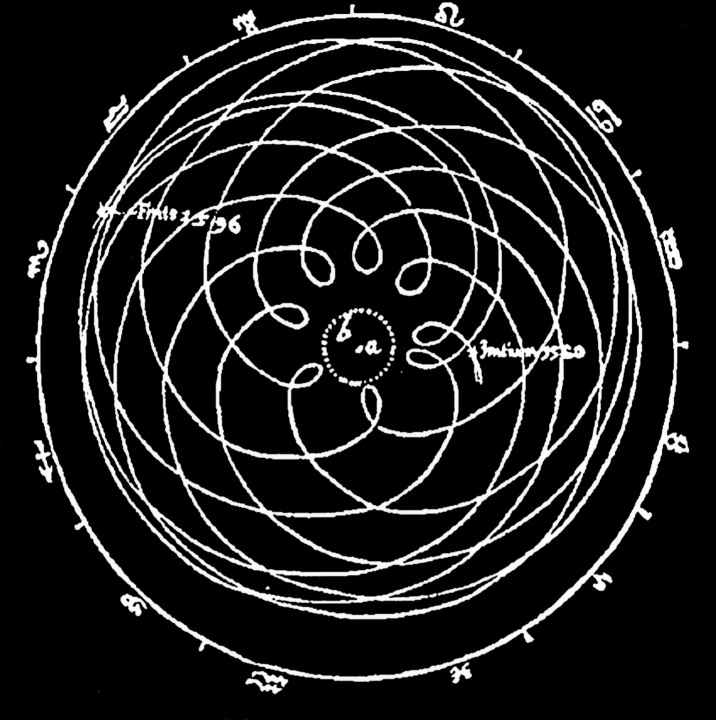

His model, published in the 1543 book *De Revolutionibus Orbium Coelestium (On the Revolutions of the Heavenly Spheres)* was still flawed as it retained the notion of spheres and perfect circular motion at uniform speed. Copernicus continued to rely on epicycles to fine-tune the apparent motion of the planets, and Mars in particular.

It was only in 1609 that German mathematician Johannes Kepler proposed a new system in which planetary orbits were not perfect circles. Based on meticulous observations of Mars by his one-time employer, Danish astronomer Tycho Brahe, Kepler argued that orbits were ellipses (stretched circles), and that planets altered their speed depending on their distance from the Sun.

**ABOVE** This ornate diagram from the *Astronomia Nova* shows the apparent position and distance of Mars relative to Earth over several orbits. Elliptical orbits proved to be the key to understanding how Mars could vary its brightness and speed in the sky, and also execute its elegant retrograde loops.

By breaking free of the old models that relied on imaginary crystalline spheres, Kepler's three "laws of planetary motion" described exactly how the planets move for the first time, and paved the way for Isaac Newton's revolutionary theory of universal gravitation (published in 1687), which would explain just why such orbits arise.

## Mars Through the Telescope

Kepler's breakthroughs coincided almost exactly with the invention and spread of the first primitive telescopes—optical instruments that used a pair of lenses to create a magnified image of distant objects. Italian physicist Galileo Galilei was one of the first people to train a telescope on the sky and, although his device was primitive by today's standards, it helped him to discover the satellites orbiting Jupiter and the Moon-like phases of Venus. Observations like these confirmed the theories of Copernicus and Kepler.

Mars, being more distant than Venus and far smaller than Jupiter, presented a much more challenging target. Finding no detectable surface features, Galileo attempted instead to observe the Red Planet's phases. It might seem odd that Mars should exhibit phases. Unlike the Moon and inner planets, where the changing visibility of dark and sunlit hemispheres creates the familiar phase patterns, Mars orbits beyond Earth, so we only ever see its sunlit face. In fact, Mars orbits close enough to Earth that, in the right configuration (known as quadrature), part of its dark side becomes visible and the planet takes on an appearance similar to a gibbous moon—a slightly less-than-full disk.

## The Dawn of Martian Geography

The weakness of Galileo's instruments left him uncertain as to whether he had detected Martian phases, but they were certainly observed in detail by Polish astronomer Johannes Hevelius in 1645. By this time telescope technology had improved considerably, and the significance of oppositions as the best time for observing Mars had also been recognized.

The earliest known report of true surface features observed on the planet came from an Italian Jesuit priest named Daniello Bartoli, who recorded two dark patches in December 1644. Particularly close oppositions in the 1650s revealed further features, but it was not until the end of that decade that the true science of Martian geography began.

An early breakthrough came with the observations of Christiaan Huygens, a Dutch scientist and instrument-maker famed for his invention of the pendulum clock and his description of the rings of Saturn.

In November 1659, Huygens used a telescope that he had made himself to observe a dark V-shaped mark on the disk of Mars—the feature now known as Syrtis Major. By tracking the dark shape over several nights, he was able to conclude that Mars was spinning on its axis in a period close to 24 hours.

Huygens never published these observations, and when the great Italian observer Giovanni Domenico Cassini carried out a similar study in 1666, he calculated a Martian rotation period of 24 hours 40 minutes—that's just 3 minutes more than the modern value.

## The London Aerial Telescope

At around the same time, Robert Hooke in London used a truly enormous "aerial telescope" to make his own observations of Mars. The telescope consisted of lenses that were separated by a distance of 36 feet (11 m). This allowed Hooke to observe Mars at around the same size as a full Moon viewed with the naked eye.

Despite severe problems with atmospheric turbulence, Hooke persevered and was able to produce some of the first maps of features on the surface of Mars.

It is Huygens and Cassini, however, who are usually given credit for discovering another type of Martian landmark—the two large white spots that mark the planet's poles. Cassini recorded seeing a brighter area at the planet's southern pole in 1666, while Huygens spotted a fuzzy white spot at the north pole in 1672. The discovery of these polar caps marked another step in our understanding of Mars as a planet similar to our own.

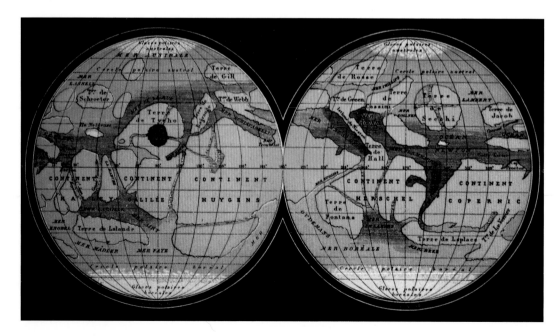

**LEFT** One of several maps of Mars compiled by the French astronomer Camille Flammarion in the late 19th century. Flammarion, an enthusiastic supporter of Martian habitability, named his continents and "oceans" after a range of famous astronomers and other scientists.

**RIGHT** A 1666 page from the *Philosophical Transactions* of the Royal Society combines early observations of Mars and other planets by Giovanni Cassini and Robert Hooke, two of the finest astronomers of their time.

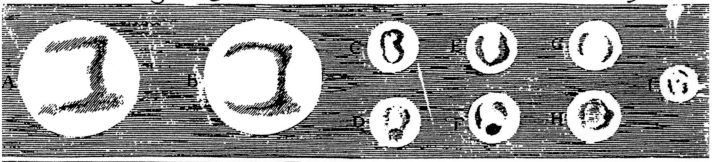

## The Obseruation of Iupiter.

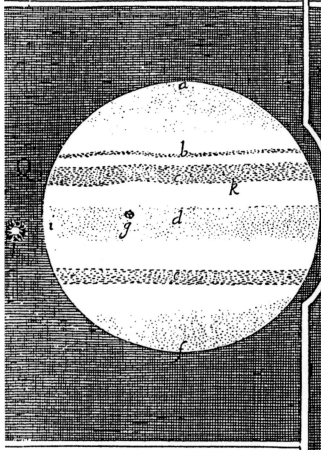

## The Figurs of y.ͤ Italian Obseruations.

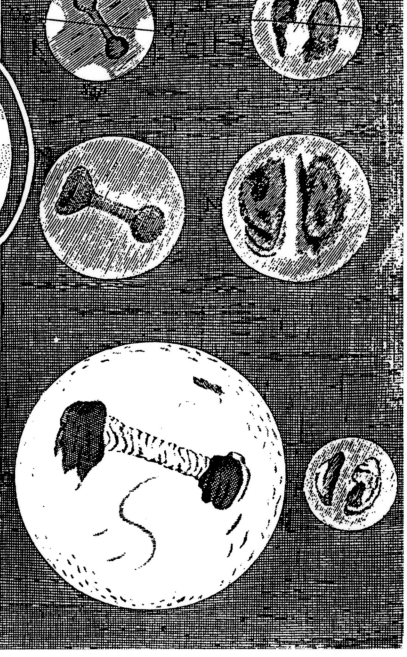

## The late Observ of Saturne.

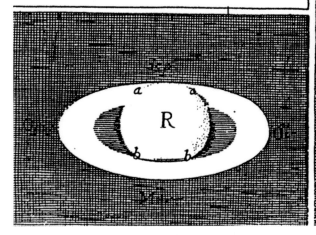

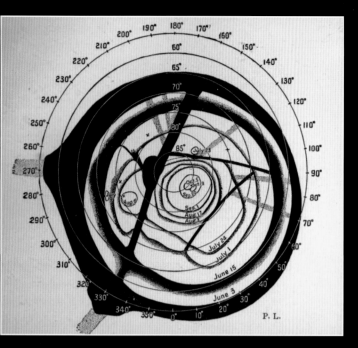

ABOVE In 1894, astronomer and Martian life enthusiast Percival Lowell systematically tracked the changing extent of the Red Planet's southern ice cap, producing this striking chart.

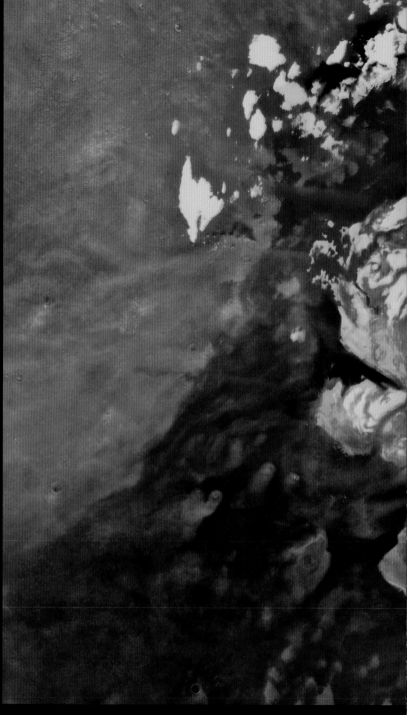

## Seasons of Mars

The first truly methodical, long-term attempt at monitoring Mars was made in the early 18th century by Cassini's nephew and assistant at the Paris Observatory, Giacomo Maraldi. He not only identified new, recurrent features on the surface, and confirmed his uncle's earlier estimate of the Martian "day," but also learned a great deal more about the planet through his studies of its polar caps.

Maraldi observed Mars at each opposition up to 1719, and became convinced that many of its surface features were changing over time. This included the darker patches that we now know are susceptible to the drifting of Martian surface sands and can even be obscured by overlying weather systems. It also included the intriguing bright areas at the north and south poles.

## Mysterious White Spots

The southern spot proved easier to observe during close "perihelic" oppositions, and Maraldi thought that its extent changed considerably over the Martian year. Noting an apparent "wobble" in the spot's position through each Martian day, he also concluded that the spot was either off center at the planet's geographical south pole, or perhaps unevenly distributed around the pole.

Maraldi did not speculate further on the Martian poles, and it was not until the late 18th century that another great observer returned to the subject. German-born British astronomer William Herschel had the advantage of his own, self-built and precisely engineered reflecting telescopes, the finest of their age. It was with these instruments that he had discovered a new planet, now known as Uranus, in 1781.

That same year, Herschel concluded that the north polar spot, like the southern one, was slightly offset from the axis of rotation. During the opposition of 1783, he used precise observations of the south pole's motion to pinpoint the geographical pole.

## Polar Ice Caps

Herschel estimated that the planet's axis of rotation was tilted at 29 degrees to the plane of its orbit—an angle that is very close to Earth's own "axial tilt" of 23.5 degrees. It seemed entirely logical to Herschel, therefore, that the polar spots were likely to be ice caps similar to those found at Earth's own poles. The way that Mars tilted meant, just as it does on Earth, that the surface areas shrouded in white received the least heat from the Sun during the course of a year and were covered in ice.

A photomosaic of images from NASA's Viking orbiters in the late 1970s reveals the intricate patterns of the Martian north polar ice cap. The dark canyonlike channels are thought to be a result of prevailing winds systematically evaporating the surface ice.

With a similar tilt to Earth, it made sense that Mars should also display a similar cycle of seasons. As Mars followed its orbit, first one hemisphere and then the other was angled toward the Sun, creating warmer and cooler periods. Transitions between the Martian winter and summer in each hemisphere would create "seasons," much as we have on Earth.

Thanks to the longer Martian year, each season would last almost twice as long as its equivalent on Earth, but the idea of seasons explained why the bright spots waxed and waned in size and intensity between one opposition and the next.

## Dry Ice and Snow

Today, we know that Herschel's ideas were substantially correct. Mars does have polar ice caps, although a great deal of the visible surface of the Martian caps is composed not of water ice, but of frozen carbon dioxide—"dry ice."

During spring, much of this thick ice transforms directly from solid to gaseous form in the thin Martian atmosphere (a process called sublimation), while in the fall, as the pole cools, carbon dioxide either snows back out of the atmosphere, or simply condenses to form frost.

More important than the scientific details at the time, however, was the fact that Herschel's work demonstrated striking similarities between Mars and Earth, igniting a wave of interest in the Red Planet that would last for more than a century.

sketches, instead coming to the conclusion that Mars was in a state of constant and frustrating flux.

## Lands and Seas

During the opposition of 1830, two other German astronomers, Johann Heinrich Mädler and Wilhelm Beer, set out to create the first complete map of Mars. By calibrating the features they observed to the position of a distinctive dark spot that lay just south of the Martian equator, Mädler and Beer were able to determine that the bright and dark patches on the surface were mostly permanent and consistent features.

After a decade of work, their map was published in 1840. It had various features that were simply designated with letters and it was not until the 1860s that astronomers started to give these areas more picturesque names, often echoing the convenient way that "seas" and "lands" observed on Earth's Moon had been identified.

Mädler and Beer's reference point, for example, was named Sinus Meridiani (Meridian Bay) by French astronomer Camille Flammarion. Subsequently used in other mapping projects, it became the marker for the Martian prime meridian. This is the line marking 0° longitude—equivalent to the Greenwich Meridian on Earth.

## Mapping Mars

Herschel's breakthroughs in the manufacture of telescopes were the first of many innovations that helped to make the 19th century a golden age for Earth-bound observers, and Martian geographers in particular. Herschel actually paid little attention to the dark markings of lower Martian latitudes but his contemporary, Johannes Schröter, took far more interest. Renowned for his close studies of the Moon, Schröter's patient studies produced a series of detailed drawings. Curiously, Schröter seemed unable to recognize the continuity of some large-scale features between his

TOP One of Percival Lowell's most detailed maps charts the canals as he reportedly saw them in 1894. Changes in the patterns perceived from one opposition to the next were put down to Martian construction works proceeding at a terrific rate.

RIGHT Success in business gave Lowell the freedom to pursue his passion for astronomy—although he is forever associated with the canal saga, the observatory he founded at Flagstaff, Arizona, remains a leading research institution.

## Martian Canals

Over the next few decades, numerous observers turned their attention to mapping Mars in ever-increasing detail and applying their own naming schemes to its features. These cartographers included the Italian Jesuit priest Angelo Secchi, Dutch astronomer Frederik Kaiser, and England's Norman Lockyer. Secchi was responsible for two significant innovations—he produced the first color illustrations of the Red Planet, and, in 1869, he described a pair of dark lines he perceived on the Martian surface as "canali."

In Italian, the word *canali* means simply "channels," but for English readers it carries an inevitable suggestion of artificial construction. Secchi's own use of the term seems to have passed unremarked, but when another Italian astronomer, Giovanni Schiaparelli, produced his own canali–strewn maps of Mars after the 1877 opposition, the idea of "Martian canals" took hold of the public imagination.

Schiaparelli was a skilled observer, but unfortunately his name will be forever associated with the tale of the non-existent canals. He certainly wasn't alone in reporting strange linear features criss-crossing between the darker areas of the Martian surface, and while Schiaparelli did not suggest that his channels were the work of some organizing intelligence, there were plenty of others who did.

## Martian Engineers

Many were quick to seize upon the theory that the existence of canals on Mars meant that there were canal builders at work on the Red Planet—alien engineers who had created a massive network of canals criss-crossing the surface of this distant world. Were they still there? Were they still at work? And, if we could see evidence of their activities, could they see us?

At the very least, the discovery of the straight lines running along the surface of the planet must surely indicate that intelligent life had at one time existed on Mars. Perfectly straight lines, after all, do not crop up in nature on Earth. Any structure that is defined by straight lines has inevitably been designed and built by men.

ABOVE Lowell at the eyepiece of a giant 24-inch (60-cm) refracting telescope he installed at Flagstaff in 1896. Before his death in 1916, Lowell dedicated funds to search for planets beyond Neptune, leading to the discovery of Pluto in 1930.

One of the foremost promoters of the idea of intelligent Martians was Camille Flammarion, a French astronomer and sometime science-fiction writer who wrote extensively on Schiaparelli's discoveries in his 1892 book *La Planète Mars*. Flammarion in turn inspired Percival Lowell, a wealthy American businessman who devoted huge resources to his passion for astronomy.

In 1893, Lowell founded an observatory at Flagstaff, Arizona, where he began an extensive program of Mars observations that resulted in the publication of three major books over the next 15 years. Lowell argued that the Martians had built the canals in a desperate attempt to establish an irrigation network transporting water from the polar ice caps to the equatorial deserts, connecting dark patches that marked the dwindling zones of vegetation on a dying planet.

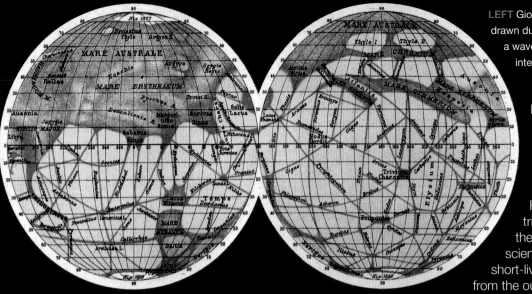

## The Coming of the Martians

In 1898, the idea of intelligent life on Mars was catapulted into the mainstream with the publication of *The War of the Worlds*, a novel by British author H. G. Wells. Taking inspiration from the works of Flammarion and Lowell, and genuine reports of strange lights from Mars a few years before, Wells wove a tale of human civilization crushed beneath the might of a Martian invasion. He had written about the possibility of Martians in a factual essay a couple of years before, and his fictional invaders were very similar—advanced, intelligent beings driven to flee their planet by the spread of deserts across its face. Rendered physically feeble by Earth's strong gravity, the tentacled Martians attack in three-legged war machines armed with devastating heat rays, defeating all military opposition only to succumb finally to common bacteria.

*The War of the Worlds* tapped into a range of concerns of the time. Invaders from space remained a minority worry, but Wells' book connected with broader concerns about hostile invasion and what a global war might mean in the industrial age. It has been reinterpreted many times, most famously in a US radio adaptation by Orson Welles in 1938, which told the story through the medium of "live" news broadcast that interrupted scheduled programming and caused widespread panic at the time. George Pal's 1953 film adaptation gave the Martians manta-ray-shaped "flying saucers," hinted at "divine intervention" in mankind's deliverance, and tapped into Cold War fears about godless communism.

The book, and Percival Lowell's own increasingly popular efforts to promote the "Martian canals," influenced many other authors in the early 20th century. *Edison's Conquest of Mars*, an unauthorized sequel published in the United States within months of the original, cast the popular American inventor Thomas Edison as the hero of a counterstrike against the Martians. Perhaps most popular of all (though less scientifically rooted) were the Martian adventure stories of Edgar Rice Burroughs (inventor of Tarzan), published between 1912 and 1942. Burroughs' hero, Civil War veteran John Carter, finds himself teleported to Mars and caught up in a

battle between the rival factions on the dying planet. Once again, canals and advanced fighting machines play important roles.

For all the fictional excitement triggered by Martian canals, their heyday as a plausible scientific claim was somewhat short-lived. They were problematic from the outset—some experienced astronomers backed up the claims of Schiaparelli and Lowell while many others did not. The truth of the matter began to emerge in 1903, when E. W. Maunder carried out an experiment. He asked a group of schoolboys to reproduce a drawing of Mars arranged to mimic the view through a telescope, and found that many spontaneously added straight lines between the darker areas. The idea that the canals were an illusion was further confirmed during the 1909 opposition of Mars, when detailed photographs of the planet were produced by astronomers at the Pic du Midi Observatory in southern France. Widespread acceptance of the "illusion" explanation did not prevent Lowell from pursuing his claims of intelligent life on Mars up to his death in 1916.

A Martian war machine emerges from its landing site on Horsell Common in Surrey, England, laying waste to onlookers with its deadly heat ray, in an early illustration from *The War of the Worlds*.

## Martian Weather

Early telescopic observers such as Cassini mistakenly thought that Mars might have a very thick atmosphere due to the premature disappearance of stars about to pass behind it. It wasn't until around 1784 that William Herschel correctly guessed that this was simply due to the stars being lost in the glare around the planet.

Using his own, superior instruments, Herschel was able to track the faint stars all the way to the moment of "occultation" by the planet's disc, and correctly concluded that the Martian atmosphere must be very thin. Nevertheless, his observation of elusive bright and dark bands made Herschel certain that Mars did, indeed, have an atmosphere that included changeable weather systems.

The first definite Martian weather feature to be identified was the so-called "Blue Scorpion." This distinctive, V-shaped cloud pattern was first spotted by Angelo Secchi in 1858. Other observers soon confirmed the phenomenon as a seasonal feature that developed over the prominent dark plain of Syrtis Major at a time when the northern hemisphere was moving from spring into summer. The cloud turns the underlying markings noticeably blue and is now known, rather less poetically, as the Syrtis Blue Cloud.

## Giant Dust Storms

As instruments continued to improve through the late 19th and early 20th centuries, other geographically restricted, seasonal clouds were discovered. A huge ring-shaped formation that became known as "the repeating northern annular cloud" appears at mid-northern latitudes on summer mornings. The cloud briefly resembles a terrestrial cyclone until it dissipates once again each afternoon.

Sometimes described as forming a "donut" shape, the ring measures about 1,000 miles (1,600 km) across, with the hole in the middle—the inner "eye" of the cloud—around 200 miles (320 km) wide.

Rising air currents created by the volcanoes of the Tharsis region, a huge volcanic plateau located in the western hemisphere near the equator, produce a different kind of cloud. The Tharsis clouds consist of water ice and assume a bright, "W" shape but follow a similar daily cycle to the circular northern annular cloud. Forming at high altitude overnight, the Tharsis clouds descend to the ground by morning, and smaller clouds form in spring and summer over the major volcanoes—the largest known volcanoes anywhere in our solar system—during the afternoon.

The most famous Martian weather systems, however, are not white or blue clouds, but the vast dust storms that periodically scour the planet's surface. The color of these storms makes them hard to identify, and the first person to suspect their presence was Schiaparelli, who, around the 1877 opposition, realized that certain features on the planet came and went from week to week thanks to some obscuring cover.

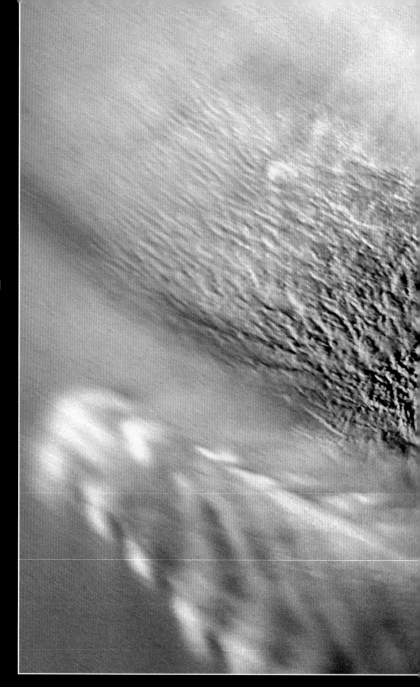

ABOVE This image captured by Mars Reconnaissance Orbiter in November 2007 shows a small dust storm (orange) close to the Red Planet's north polar ice cap. Similar small storms develop along the edge of each cap during late winter thanks to daily temperature changes.

## Danger to Space Probes

During the opposition of 1909, the clouds were once again particularly prominent, while in 1911, almost the entire Martian disk was shrouded in yellow dust. Greek astronomer Eugène Antoniadi concluded that periodic weather events lofted huge amounts of dust into the Martian atmosphere, from where it could take many months to settle back to the ground.

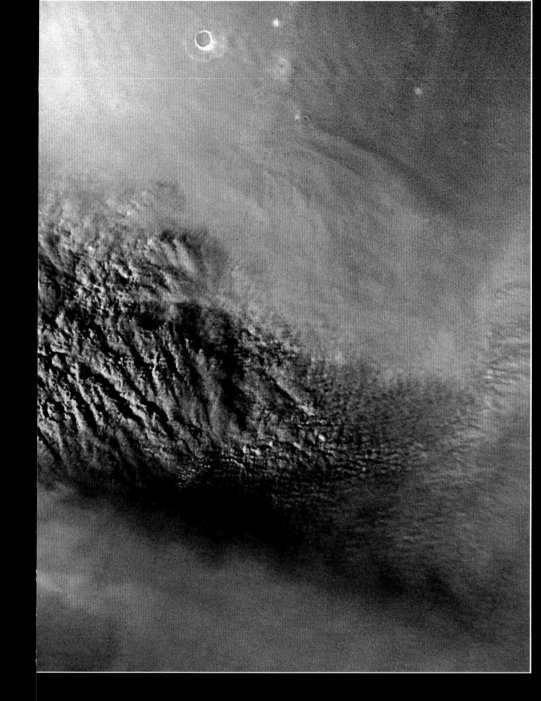

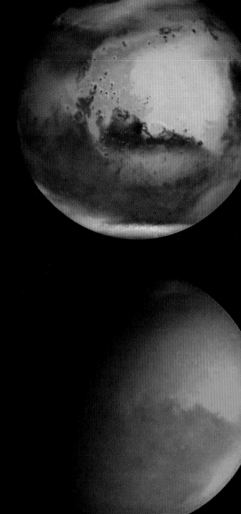

Antoniadi noticed that the storms always seemed to be at
their strongest in southern summer, when Mars is at perihelion
(closest to the Sun), and correctly concluded that this was
because the increased heat from the Sun helped create
stronger winds in the southern hemisphere.

Annoyingly for Earth-based observers, this means that storms
are particularly common during the closest approaches of
Mars to Earth, and often hamper what should be our best
views of the planet.

When space probes and other scientific equipment were first
sent to Mars, there was some concern that the violent storms

instruments deployed to investigate conditions on the plane[t.]
Fortunately, these fears have proved to be almost entirely
unfounded, mainly due to the composition of the dust.

The first Martian landers discovered that the sands of Mars
are much finer than those on Earth, and in the low-pressure
Martian atmosphere they can do little damage even when
traveling at high speed.

It is the dust's very lack of weight, however, that ensures it
remains high in the atmosphere long after the winds that firs[t]
stirred it have abated, persisting for months before the skies
clear, and even then still lending the Martian atmosphere its

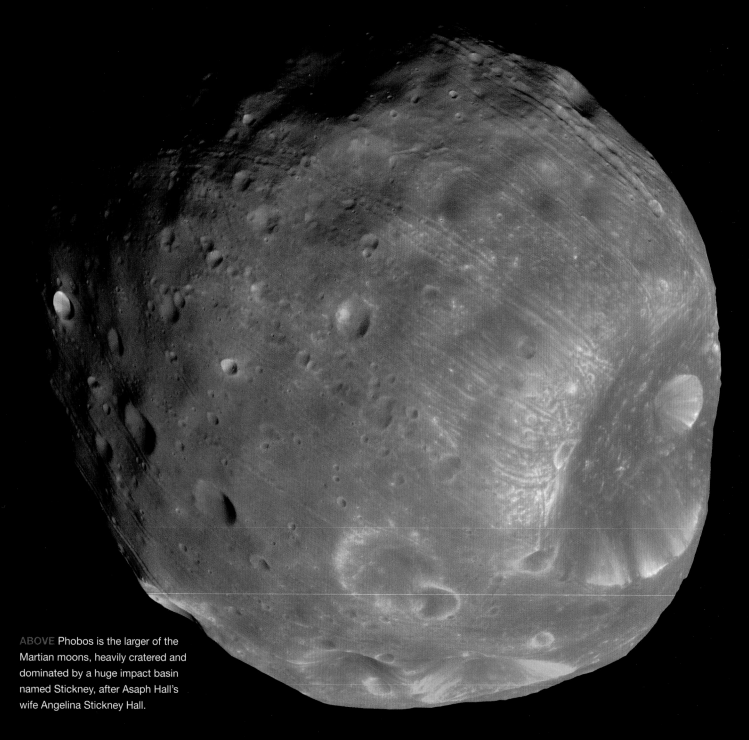

ABOVE Phobos is the larger of the Martian moons, heavily cratered and dominated by a huge impact basin named Stickney, after Asaph Hall's wife Angelina Stickney Hall.

## Missing Moons

Ever since Galileo turned his telescope on Jupiter and discovered four moons in orbit around it, astronomers puzzled over the apparent absence of satellites around Mars. By the mid-19th century, the question of Martian moons was becoming a pressing issue. Earth had one satellite, Jupiter and Saturn had several. Even the newly discovered Uranus and Neptune had moons of their own—so why not Mars?

While it was true that Venus and Mercury both lacked satellites, which could, perhaps, be explained by their position on the Sunward side of Earth, Mars had no such excuse. With so many telescopes trained on the planet, however, why had none yet been discovered?

## The Largest Telescope

In 1877, American astronomer Asaph Hall began a deliberate search for Martian moons, using what was then the world's largest lens-based telescope at the U.S. Naval Observatory in Washington, D.C.

Hall realized that the planet's weak gravity would not be able to hold on to a satellite at any great distance. Consequently, he decided to concentrate his attention close to the Martian disk. The power of the telescope he was using would give him the best chance of locating a satellite in close orbit that might otherwise remain unnoticed against the turbulent Martian background and the planet's glare. Hall's efforts were to bear fruit surprisingly quickly, although he almost lost patience

Beginning the quest in early August, at first, like many astronomers before him, Hall could see no moons. Frustrated, he considered giving up but was encouraged to persevere by his wife, Angelina. The following night, on August 11, he found the first moon.

## Deimos and Phobos

Finding a small moon orbiting Mars convinced Hall to continue his search and, six nights later, he found a second, larger satellite even closer to the planet. The moons were given the names Deimos, for the smaller, outer moon, and Phobos for the larger moon in closer orbit of the planet.

The names come from Greek mythology, Deimos and Phobos being the sons of Ares, who was the god of war. In Greek, the name Deimos is taken to mean panic, terror, or flight, while Phobos means fear. Roman gods were closely associated with the earlier Greek deities and the Roman equivalent of Ares was Mars.

Although the Greeks knew them as the sons of Ares, Deimos and Phobos are also said to be the names of the horses that drew Mars' chariot.

## The Smallest Moons

Compared to Earth's Moon or any of the giant-planet satellites then known, the new Martian moons seemed tiny. Earth's Moon, although nowhere near the largest planetary satellite in our solar system, has a circumference around its equator of 6,786 miles (10,921 km). Both of Mars' moons were found to be much smaller, irregular lumps of rock and debris.

We now know that Phobos, the innermost of the Mars moons, is by far the larger of the two with dimensions of roughly 17x14x12 miles (27x22x19 km). Deimos measures some 6.5x7.5x10 miles (11x12x16 km) and is even more irregular in shape than Phobos.

The close proximity of both moons to Mars gives them rapid orbits. Deimos circles the planet in 30 hours and 21 minutes at an average distance from the surface of 12,500 miles (20,000 km). Phobos is only 3,750 miles (6,000 km) out and completes its circuit in just 7 hours and 40 minutes. That rate of orbit outpaces Mars' daily rotation so that the moon goes backward across Martian skies, rising in the west and setting in the east less than six hours later.

Working out from where the moons came has presented planetary geologists with a real puzzle.

## Origins of the Moons

Initially, Phobos and Deimos were assumed to have been stray asteroids that drifted into the influence of Mars' gravitational field and were captured into orbit. Analysis of light reflecting off the surfaces of the moons suggests that they may have a similar composition to some asteroids, providing evidence for the "gravity trap" theory.

Yet some scientists argue that the odds against one asteroid, let alone two, approaching Mars at just the right trajectory to avoid either flying straight by or smashing directly into the surface are literally astronomical. An alternative theory is that the two tiny satellites may have formed in the same way as Earth's own Moon. Around 4.5 billion years ago, an asteroid approximately the size of Mars is thought to have crashed into Earth. The impact threw out a huge cloud of debris that eventually came together to form the Moon. Deimos and Phobos are believed to have been created by major asteroid impacts on Mars.

If this second theory is correct, then Phobos and Deimos might be the last survivors from a ring of small moons that once encircled Mars. Tidal forces caused by Martian gravity would have gradually drawn the closest of them in toward the planet.

Phobos is certainly doomed to a similar fate. Its orbit is gradually deteriorating and eventually tidal stresses will see it shatter into pieces that will become a short-lived ring around Mars before the fragments rain down onto the equatorial region, forming a chain of fresh craters.

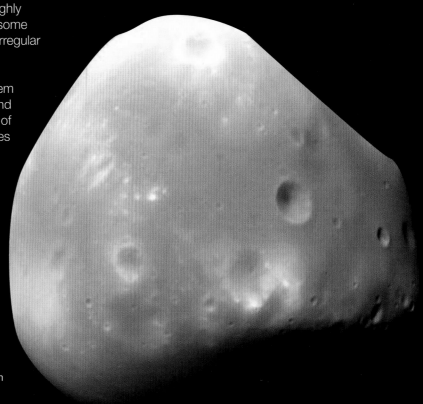

RIGHT Deimos has a much smoother appearance than Phobos: its surface has been blanketed by dust and debris from impacts, hiding the detail of the object beneath.

## A Dead Planet?

The first half of the 20th century saw significant advances in our knowledge of Mars, but for those who might have hoped that Mars would prove to be habitable, capable of colonization as an "overspill" planet to take some of Earth's mushrooming population, there was only disappointment in store. Almost all of the facts pushed the general picture away from that of a habitable planet, and toward a harsh, cold, and arid world incapable of supporting life.

The dismissal of the Martian canals was an early blow, with both photographic evidence and improved observations indicating that they were an illusion. Even before that, however, American astronomer William W. Campbell had cast doubt on ideas that Mars had a wet and relatively Earthlike atmosphere through his spectroscopic measurements of 1894.

These measurements were made by splitting the planet's light into a rainbowlike spectrum and analyzing the strength of different colors. This indicated which chemicals were present in the atmosphere on Mars. Campbell's findings, along with his own research, inspired no less a figure than Alfred Russel Wallace (co-discoverer, with Charles Darwin, of the theory of evolution by natural selection) to address the problematic Red Planet in his 1907 book *Is Mars Habitable?* The answer to that question, according to Wallace, was a resounding "No."

Despite Wallace's support, Campbell's discouraging findings were dismissed by many at the time, only to be confirmed beyond a doubt by fellow American, Walter S. Adams, in 1925. Adams proved that the Martian atmosphere contained little if any oxygen—another crucial blow to hopes for life.

## Chilly Temperatures

Also in the mid-1920s, astronomers used the enormous 100-inch (2.54-m) Hooker Telescope at the Mount Wilson Observatory in California to make direct measurements of the temperature on Mars for the first time. They found these ranged from around -94°F (-70°C) near the poles, to perhaps 45°F (7°C) at the equator in daytime. When others carried out similar measurements for the nighttime temperature, they found these plunged to around -130°F (-90°C) across the entire night side.

These temperatures were lower than those predicted by Lowell and other advocates of Martian life, but matched the chilly predictions of Wallace. We now know that daytime temperatures can reach up to 68°F (20°C) in the summer at the equator but fall as low as -243°F (-153°C) at the poles.

It became clear during the 1920s that the Martian atmosphere was, at best, tenuous. In 1929, French astronomer Bernard Lyot measured the way the gases in Mars' atmosphere scatter sunlight, concluding that they were less than 7 percent as thick as those in Earth's atmosphere.

In 1947, Dutch-born American Gerard Kuiper made spectroscopic measurements confirming the atmosphere was dominated not by oxygen but by carbon dioxide, which is toxic to animal life. The following year, U.S. meteorologist Seymour Hess showed that the clouds seen in the Martian atmosphere were probably a result of small amounts of water vapor condensing under very low atmospheric pressures—just 1 millibar, or about 1/1,000th of the average atmospheric pressure on Earth. In the space of a few decades, Mars had gone from being a potential abode of life to a dead world, albeit one about which we knew more than ever before.

Eugène Antoniadi, whose observations helped to discredit the canals, created some of the most detailed Martian maps so far, and named many of the large-scale "albedo features" (patches of light and dark) that would later be acknowledged by the International Astronomical Union in the first official gazetteers of the planet.

**MAIN IMAGE** A modern artist's impression of Mars depicts it as a cold, dry, and dusty desert with a thin atmosphere, incapable of sustaining life as we know it.

**BELOW** Eugène M. Antoniadi's maps of Mars set the "gold standard" for observations up until the dawn of the Space Age, and many Martian features still bear the names that he assigned to them in his early-20th-century maps.

## Mars in the Space Age

The era of space exploration that began in the second half of the 20th century was to transform entirely our picture of Mars, mostly through images and data sent back by automated space probes. While the work of these robot missions is covered in greater detail in the following chapters, the emergence of new technology that allowed more accurate and detailed observations from Earth has also proved important.

Perhaps the most iconic of all Earth-based observatories is the Hubble Space Telescope, a satellite launched into a low orbit above the bulk of our planet's atmosphere in 1990. Hubble's advanced optics have given us unprecedented views of objects throughout the Universe, ranging from distant galaxies to the planets of our solar system, while the ability to service, replace, and upgrade its cameras and other instruments has delivered a wealth of scientific data over two and a half decades of operation.

Since its launch, Hubble has observed Mars at every opposition. This proved invaluable during the 1990s, when there were no operational space probes in orbit around Mars, and even today it provides a valuable complement to the more tightly focused views of orbital surveys. Hubble has, for instance, been used to track the large-scale weather systems covering entire hemispheres, and to monitor seasonal changes, such as the growth and retreat of the ice caps.

Even ground-based observatories still have a role to play in the Space-Age exploration of Mars. Advances in technology allow the building of telescopes that are larger than ever before, and capable of delivering Hubble-quality

ABOVE The first image of Mars captured by Hubble in 1991 showed a huge advance on previous Earth-based observations. Subsequent upgrades to the telescope have only served to improve its performance.

images through the still atmospheres of mountain-top sites. In 2009, astronomers using the 33-foot (10-m) Keck telescopes on Hawaii and NASA's neighboring Infrared Telescope Facility found the first hints of methane in the Martian atmosphere—observations that have since been backed up by the Curiosity rover on the planet's surface. Telescopes around the world, meanwhile, turned their attention toward Mars in October 2014 as Comet Siding Spring buzzed past the planet.

Hubble's successor in orbit, the James Webb Space Telescope, is an infrared observatory with a 21-foot (6.5-m) main mirror, scheduled for launch in 2018. While its main goal will be to answer cosmological questions about the early days of the Universe, it will also give us a new view of Mars, thanks to spectroscopic instruments that can analyze the Martian atmosphere at high enough resolution to see how it is affected by the planet's surface. This may help to solve unanswered questions about the polar caps and the icy soil that surrounds them, and perhaps also help to detect short-lived flood events or low-level volcanic activity.

On Earth, astronomers and engineers are planning the next generation of ground-based telescopes, with diameters of up to 100 feet (30 m). It could be a decade or more before these giant eyes on the sky become operational—far too soon to start drawing up lists of observing targets—but it seems almost certain that, when the time comes, the Red Planet will once again be high on everyone's list of priorities.

LEFT The Hubble Space Telescope's unique location outside Earth's atmosphere gives it crystal-clear views of the planets and deep space.

# FIRST RECONNAISSANCE

As humanity took its first faltering steps in the exploration of outer space, Mars was an obvious early target. Science-fiction visions of ambitious manned landings had fired the beginnings of the Space Age but, in reality, the first close-up images of the Red Planet would be sent back by automated space probes flying past at high speeds. Their occasional visits throughout the 1960s produced flip-flopping views of the Martian environment, while hinting at intriguing geographical features. It was only in the 1970s that the first probes entered long-term orbits, and robot landers made their first touch-down on the dusty surface.

ABOVE Transported to the United States alongside many of the German V-2 "Rocket Team," Werner von Braun (center) was confined to work on military rockets until the launch of the first satellites saw the Soviets take an early lead in the space race. Called in to save the U.S. space program, von Braun hobnobbed with presidents including John F. Kennedy (right), arguing constantly for mankind's long-term future in space.

## Deadly Rockets

Although dreams of Mars as a hospitable planet with potential for abundant life faded with the discoveries of the early 20th century, Mars still loomed large in the imaginations both of science-fiction writers and the pioneers of real-life space travel. Groups of rocketry enthusiasts, often fired by the writings of Jules Verne, H. G. Wells, and even Edgar Rice Burroughs (whose take on Martian exploration was altogether more fantastical), formed in the United States, the Soviet Union, Britain, Germany, and other countries, but it was the Germans who stole a technological lead, albeit at the price of supporting the Nazi military machine.

The terrible result was the V-2, a rocket-bomb that rained destruction on Britain and northwest Europe during the final

The famous U.S. launch site at Cape Canaveral started in a small way, with the testing of rockets such as this modified German V-2 in 1950. By the 1960s, it had become an enormous launch complex and departure point for all NASA space probes to Mars.

**RIGHT** Today we take digital imaging and data transmission for granted, but early space probes relied instead on analog TV cameras. The picture was sent back one line at a time, and color data added later—this image shows an impatient researcher during the Mariner 4 flyby speeding up the process with a "paint by numbers" approach. Mariner 4 was launched on November 28, 1964, and journeyed for 228 days to the Red Planet, providing the first close-range images of Mars.

months of World War II. In the war's aftermath, the Soviet Union and United States scrambled to get hold of German rocket technology and the brains behind it, often turning a blind eye to the potential crimes of those involved. Both sides in the burgeoning Cold War saw rockets as a potential delivery system for the nuclear weapons that would hold the future balance of global power, but the engineers and enthusiasts on both sides who were driving rocket research forward were swift to press their argument that space exploration would be a valuable propaganda tool.

Germany's V-2 mastermind Werner von Braun, who worked for the U.S. military after the war, wrote a series of influential articles outlining a long-term strategy for the exploration of space, culminating in a manned mission to Mars. Published in the widely read *Colliers Magazine*, they later formed the basis for a Disney television series that fired the public imagination.

## The Space Race

By the late 1950s, America was dreaming of a new "manifest destiny" in the stars, so it was an enormous shock when, in October 1957, their Soviet rivals beat them to it with the launch of the first satellite, Sputnik 1. Working in secret under the leadership of the hugely talented engineer Sergei Korolev, they were able to pull off a series of increasingly impressive space spectaculars, including flybys and crash-landings on the Moon, and eventually the first manned launches to orbit. The Americans were forced to play catch-up in what had suddenly become a space race.

Ironically, one key reason for the early Soviet advantage in space was their more primitive technology in another field. While the U.S. had successfully developed the lightweight and powerful hydrogen bomb, the Soviets were still reliant on older, heavier atomic bombs that required much bigger rockets. America's newly formed Space Agency, NASA, struggled to develop the powerful rockets needed to send space probes and manned missions beyond Earth's immediate vicinity, while the Soviets were already turning their attention to interplanetary exploration—and the Red Planet was firmly in their sights.

## First Impressions

As it turned out, the initial interplanetary launches also marked the first major setbacks to the Soviet space program. Two launches were planned to take advantage of the Mars opposition (close approach) of late 1960, but both missions (codenamed "Mars 1M") failed even to reach Earth orbit. Problems with the fuel lines starved the engines, which could not then achieve the required thrust, and the spacecraft crashed back to Earth, although one did reach an altitude of around 75 miles (120 km).

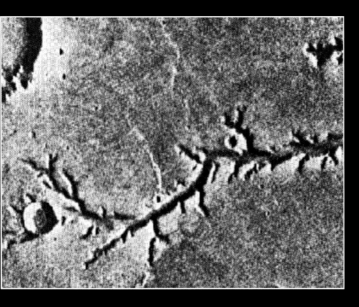

ABOVE One of Mariner 9's most important discoveries was Nirgal Vallis, a long and sinuous valley with "tributaries" that (unlike the larger Valles Marineris) could not have formed by faulting alone, and therefore provided some of the best evidence for past Martian rivers.

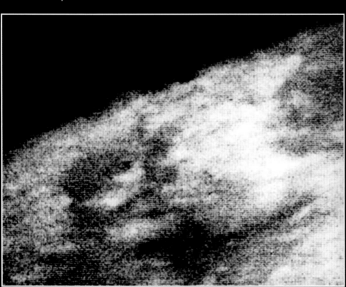

ABOVE The first ever close-up image of Mars was taken by Mariner 4. It showed an area about 185 miles (300 km) across and 750 miles (1,200 km) deep, from the bottom of the frame to the distant limb (edge of the planet).

The changing alignment of Earth and Mars then dictated a two-year wait before further attempts could be made. This time, the Soviets launched three probes—two intended flybys and a lander.

One of the flybys actually managed a moderately close 120,000-mile (190,000-km) flyby of Mars, but did not send back any data due to a communications failure en route. The other flyby probe, Mars 2MV-4 No.1, reached low-Earth orbit before its main engine exploded, destroying the spacecraft. The would-be lander also reached orbit, but an engine failure meant that it went no farther. Its orbit decayed and it broke up on re-entry the following day.

## Triumph for NASA

By 1964, NASA had begun to recover from its faltering start in the space race, putting plans into place that would ultimately see a triumphant manned lunar landing in 1969. It was also ready to aim for Mars, with a strategy that involved launching pairs of identical missions at each opposition.

The ingenuity of this tactic was demonstrated almost immediately by the Mariner 3 and 4 flyby missions. Mariner 3 hit trouble when part of the casing that enclosed the craft during launch failed to open properly once it was in outer space. Unable to deploy its solar panels, Mariner 3 could not recharge its batteries and ran out of power. It is now derelict, floating silently in orbit around the Sun.

Mariner 4, on the other hand, had a successful interplanetary trip. Launched in November 1964, Mariner 4 maintained contact throughout its outward journey and relayed close-up pictures of Mars from within 6,000 miles (10,000 km) of the surface in July 1965. These were the first ever photographs of the surface of another planet to be transmitted back to Earth from deep space.

The Soviets once again lost both of their planned missions in the 1964 launch window, although the first of these was actually an attempt to send a lander to Venus. Breakdowns in communications meant that neither craft could relay any data back to Earth. Neither was it possible to control the probes properly, although the Venus probe got to within 60,000 miles (100,000 km) of its target. The Mars mission did even better, despite limping along on half power due to the loss of one of its solar panels, passing the planet at a distance of less than 1,000 miles (1,600 km).

## Cosmic Rays

Other priorities meant that NASA did not attempt another launch until Mariners 6 and 7 of 1969, twin missions that both made successful flybys. Mariner 6 was launched on February 24, 1969, and Mariner 7 a month later. They orbited Mars over the equator and polar regions, analyzing the atmosphere and taking over 200 photographs.

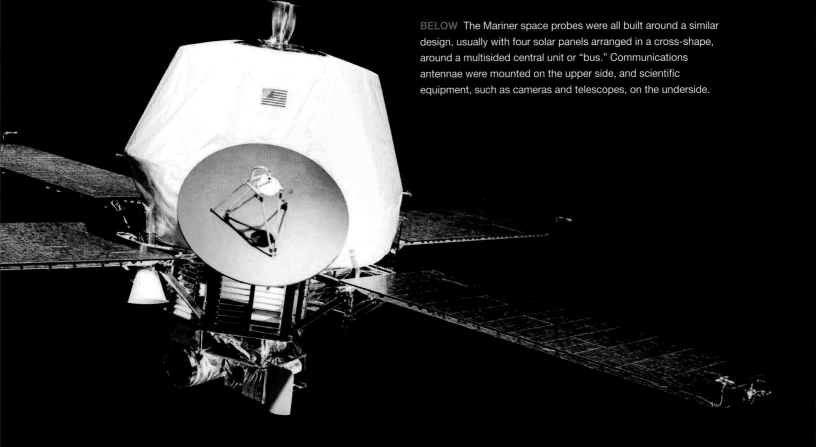

By pure coincidence, all three of the successful early Mariner missions made their closest approaches over the Red Planet's southern hemisphere, sending back pictures of the southern highland region. The results dealt a final, crushing blow to those enthusiasts who still dreamed of a Mars that might once have been habitable.

The Mariner photographs revealed a heavily cratered, arid landscape that appeared to have been shaped mostly by impacts from space. Far from being a planet like Earth, this was clearly a world not dissimilar to our own Moon.

Other measurements, meanwhile, revealed that Mars lacked the strong magnetic field required to protect it from bombardment with "cosmic rays," high-speed particles from space that can damage and disrupt living tissues. Further data confirmed the sub-zero temperature of the planet's cold, dry desert regions.

## Last Mariner to Mars

Despite these disappointments, it was clear that the next step in understanding Mars would be to put a spacecraft in long-term orbit to survey the entire planet, and this became the focus of NASA's 1971 launches.

Once again, the strategy of identical paired spacecraft paid off, as the first mission, Mariner 8, ended shortly after lift-off.

Having escaped Earth's atmosphere, a gyro failure left the craft tumbling out of control, its engine cut out and it re-entered above the Atlantic, crashing into the ocean around 350 miles (550 km) north of Puerto Rico.

Mariner 8's twin, Mariner 9, enjoyed a far more successful launch, leaving Earth in May 1971. This would be the final Mariner mission to Mars. The last Mariner probe, Mariner 10, would be destined for flybys of Mercury and Venus.

Following a five-and-a-half-month cruise, Mariner 9 entered orbit around Mars successfully on November 14, 1971. Frustratingly, the first pictures it sent back showed only a featureless globe—Mars was engulfed in one of its periodic dust storms (still the largest of its kind on record). But as the dust began to settle over the following months, features began to emerge from the haze.

Over the course of its mission, which concluded in October 1972, the spacecraft sent back more than 7,300 images, mapping 85 percent of the Martian surface from altitudes as low as 1,000 miles (1,600 km).

These included the first pictures of many well-known Martian landmarks, and early evidence that Mars had a much more intriguing and complex past than that suggested by those early flyby missions.

# Tharsis

The most prominent Martian feature of all is a vast bulge in the planet's surface, topped by several enormous volcanoes. This region, named Tharsis (from the biblical Tarshish, westernmost land of the known world) dominates the Western hemisphere of Mars. It is roughly 3,000 miles (5,000 km) across and more than four miles (7 km) higher than the average Martian "surface datum," the equivalent of Martian sea level.

Tharsis is divided into two distinct subprovinces. Its northern rise is dominated by a single low-lying volcano known as Alba Mons, and its vast network of surrounding lava flows. The larger southern rise is topped by a chain of three somewhat smaller but far taller volcanic peaks, collectively known as Tharsis Montes. Running from northeast to southwest and straddling the equator, they are individually called Ascraeus

Mons, Pavonis Mons, and Arsia Mons. The subprovince continues south until it merges with the planet's cratered southern highlands, and extends east to surround much of the enormous Valles Marineris canyon system.

The major features of Tharsis were among the first to be spotted by the Mariner 9 mission in 1971. As the global dust storm that had enveloped the planet on the probe's arrival began to subside, the rise was the first region to emerge from the haze, revealing a distinctive chain of three obvious spots. As the dust cleared further, these transformed into enormous shield volcanoes—vast heaps of layered volcanic lava with collapsed central pits known as calderas. Each has a diameter measured in the hundreds of miles, while the tallest, Ascraeus Mons, rises to over 9 miles (15 km) above the surrounding plateau, or 11 miles (18 km) above the surface datum, dwarfing Earth's largest volcano, Hawaii's Mauna Loa. Though even larger in extent, Alba Mons is somewhat less conspicuous. It was not discovered until the following year, when NASA mission scientists began to realize the full, enormous scale of the Tharsis region.

BELOW Parallel canyons and pits running along the flanks of Alba Mons may be some of the best evidence for relatively recent volcanic activity. Known as catenae and fossae, they seem to reveal the path of subterranean lava channels.

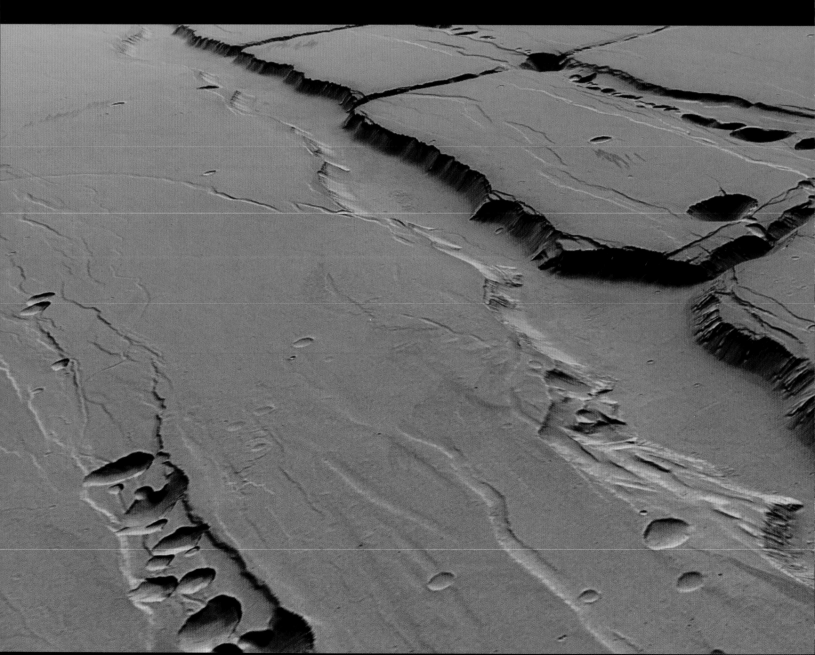

The obvious straight-line chain formed by the Tharsis shield volcanoes is irresistibly reminiscent of Earth's own Hawaiian volcano chain, and so it's tempting to seek similar explanations. The Hawaiian volcano chain is thought to lie over a "hot spot" in Earth's mantle, a plume of particularly high-temperature rock rising from deep within our planet's interior and heating the underside of the rocky crust to create a huge reservoir of molten magma that forces its way to the surface. As the thin oceanic plate of the Pacific Ocean moves across the hot spot (pushed by the forces of plate tectonics —what used to be known as "continental drift"), the location of activity shifts over millions of years. This gives rise to brand new volcanoes, even as the magma supply to the old ones falters and dies.

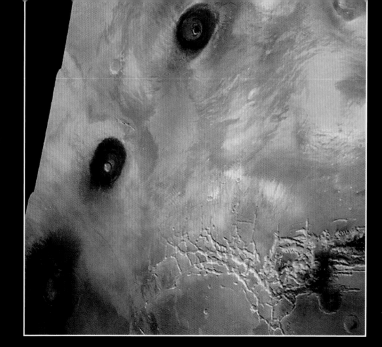

**RIGHT** A mosaic image from the Viking 1 orbiter shows the three dark domes of Tharsis Montes—Arsia, Pavonis, and Ascraeus Montes (from top to bottom). The maze-like canyons of Noctis Labyrinthus at the end of the Valles Marineris can be seen at lower right, with another volcanic shield, Tharsis Tholus, above it.

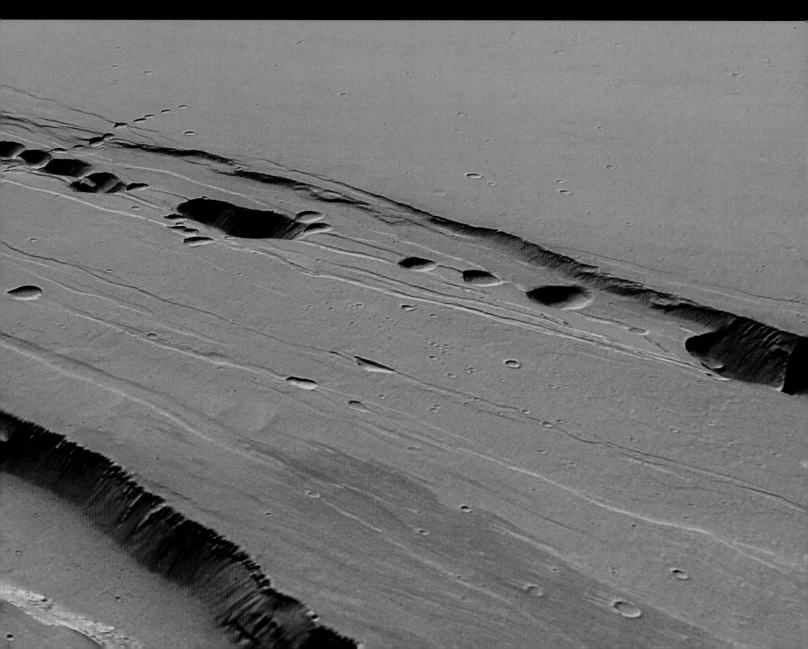

N →

## Olympus Mons

Lying just to the northwest of the Tharsis Rise, Olympus Mons is an even more impressive volcano, and one of the few Martian features that was seen from Earth before the Space Age. In fact, it was first spotted by Giovanni Schiaparelli as early as 1879, who named the bright patch on the Martian surface Nix Olympica ("The Snows of Olympus"), after the mountain home of the gods in Greek mythology. Schiaparelli predicted that it might be a mountain on the basis that it was one of the features least affected by global dust storms, and indeed the peak was one of the first to emerge from the Martian haze, alongside the Tharsis Montes, shortly after Mariner 9 arrived in orbit in 1971.

Olympus Mons is the largest mountain in the solar system — an enormous shield volcano that rises to 13 miles (21 km) above the surface datum. Because the surrounding plains of Amazonis Planitia are some of the lower-lying parts of the Martian surface, perhaps depressed by the volcano's

enormous bulk, a complete climb from the foot to the peak would ascend through an even more impressive 16 miles (26 km). However, much of the climb would be barely noticeable thanks to the mountain's enormous extent — it has a diameter of 370 miles (600 km) — and, aside from an impressive surrounding frill or escarpment of towering cliffs, much of the rise is surprisingly shallow.

At the mountain's peak is an impressive crater complex known as a caldera, some 50 miles (80 km) across at its widest. This consists of six separate craters but, although there is some evidence for lava flowing on their floors, these were never the focus of the mountain's volcanic activity. Like its neighbors in Tharsis, and the largest volcanoes on Earth, Olympus is a volcanic shield, a shallow dome formed as layer upon layer of lava poured onto the planet's surface over millions of years from an underlying magma reservoir.

The bulk of these eruptions would have come not from the peak, but from cracks along the flanks of the volcano, and as more and more lava poured out and solidified, the volcano rose ever higher. Eventually, once its magma reserves were exhausted, the subterranean chamber emptied, and upward pressure to support the mountain's enormous overlying bulk was withdrawn. The highest parts then slumped down, creating the ring-shaped calderas that have enormous cliffs up to 2 miles (3 km) high.

Each caldera results from the ending of a specific phase of volcanic activity and withdrawal of magma, so geologists believe that the volcano went through at least six phases of activity. By counting the accumulation impact craters on their surfaces (and assuming such events happen effectively at a random and uniform rate), geologists have estimated the age of various calderas to between 350 million and 150 million years. This makes Olympus the youngest of the major Martian volcanoes, and "crater counts" on the volcano's northwest flank show some

surfaces are even younger, indicating lava flows perhaps as recently as 2 million years ago.

Surrounding the escarpment, and particularly prominent to its northwest, is an apron of chaotic grooved terrain known as the Olympus Mons aureole. This covers an even larger area than the volcano itself, reaching up to 465 miles (750 km) beyond its outer cliffs, and is thought to have formed as a result of massive landslips in the distant past.

**BELOW** This stunning overview of Olympus Mons was compiled from images sent back by the Viking 1 orbiter. It shows the steep escarpment that surrounds the volcano around most of its rim.

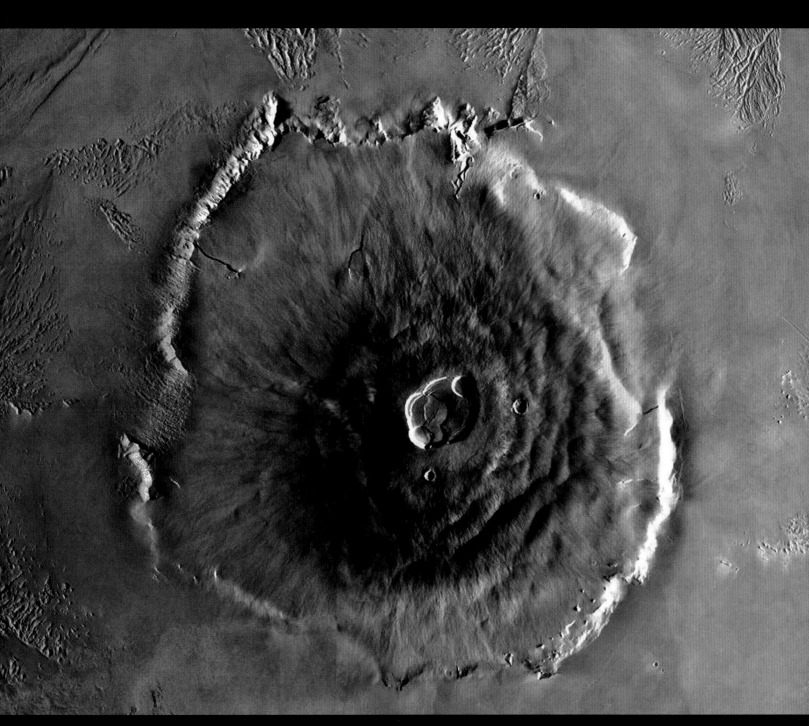

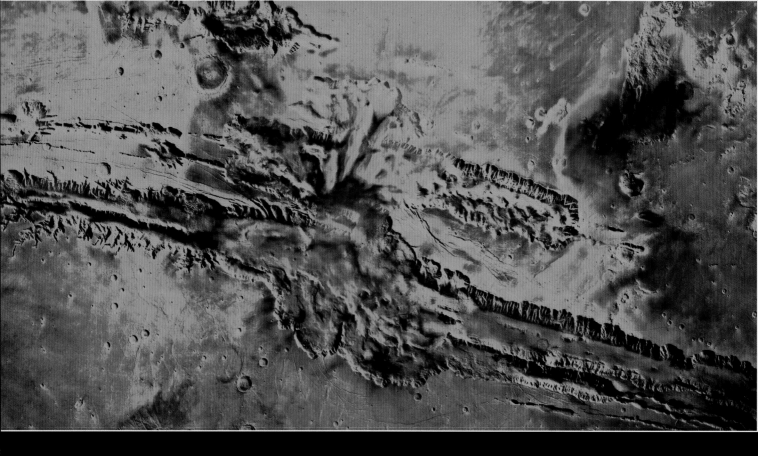

A Viking mosaic view of Valles Marineris reveals multiple parallel channels extending to the east and west of a much broader central depression. From end to end, the canyon system is more than 2,500 miles (4,000 km) long.

## Valles Marineris

After its volcanoes, perhaps the best-known and most spectacular feature on Mars is an enormous canyon system that runs in a roughly east-west direction just south of the planet's equator. Discovered in Mariner 9 photographs from 1971, it was subsequently named the Mariner Valleys (Valles Marineris) in honor of the mission.

Emerging from the Tharsis rise to the southeast of the major volcanoes, the canyon begins in a region of jumbled terrain with the picturesque name of Noctis Labyrinthus. From here, it runs for more than 2,500 miles (4,000 km)—equivalent to a valley on Earth running the width of the entire continental United States.

Broadening out into a complex series of parallel trenches and collapsed valley floor that are some 370 miles (600 km) across at their widest point, Valles Marineris emerges through chaotic terrain into a low-lying plain called Chryse Planitia. Up to 6 miles (10 km) deep in places, the valley system

dwarfs Earth's own Grand Canyon (the most obvious comparison) in every way. The only truly comparable trenches elsewhere in the solar system are the rift valleys between Earth's continental plates and Baltis Vallis, a huge volcanic lava channel on Venus.

Although there is plenty of evidence that water (or some other fluid) once flowed through the canyon, it was certainly not responsible for its original formation. Instead, the Valles seems to have begun life as a fracture in the Martian crust that split apart and widened over millions of years.

## Martian Plate Tectonics

Usually, this fracture is attributed to the high pressures in the crust produced by the creation of the Tharsis rise, but evidence has recently been put forward to support a more Earthlike origin. Some argue that the opposite sides of the valley show signs of an east-west "slip" of more than 90 miles (150 km) in respect of one another. This is the kind of movement caused by shifting tectonic plates, as seen on Earth in geological features such as the San Andreas fault.

If further studies back up this idea, it would not only alter the way we interpret the Valles Marineris, but also change our whole understanding of early Martian geology. Did the Red Planet once begin to develop tectonic plates, only for

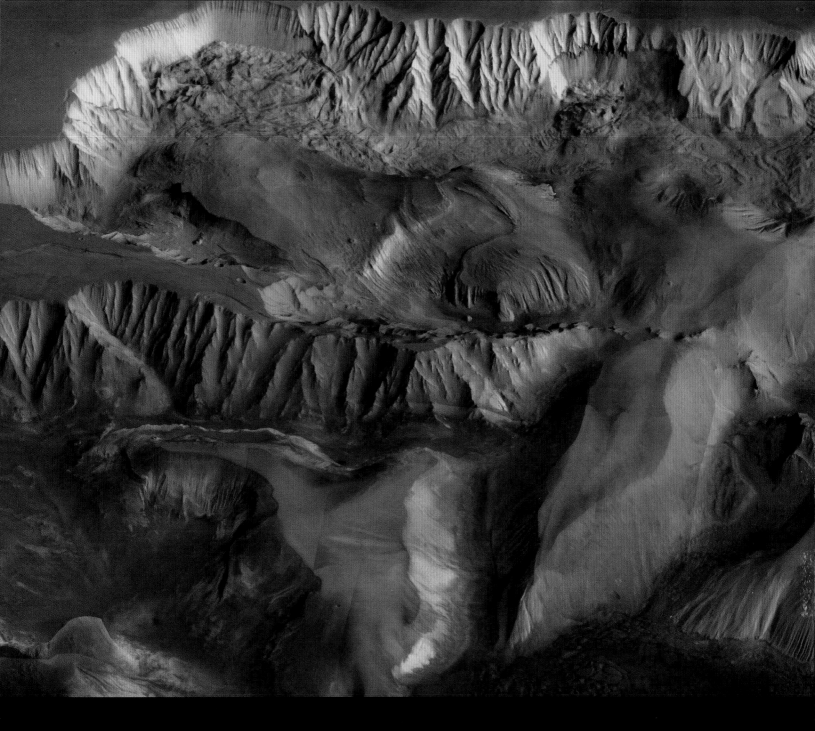

the process to grind to a halt? And what could have happened to end the formation of the plates?

## Torrential Floods

However the valley actually began its life, it seems clear that erosion has played the greatest role in shaping it since. At some point, frozen water deposits in the Tharsis soils melted, most likely due to heat from volcanic activity, causing widespread subsidence in the Noctis Labyrinthus. Torrential floods coursed along the canyon, reshaping it before flowing out into Chryse Planitia where they created teardrop-shaped islands not dissimilar to "scablands" created by catastrophic floods on Earth near the end of the last ice age.

There are also signs of water flowing through the canyon at a more gentle pace, and perhaps even forming substantial standing lakes. As the Martian atmosphere thinned and

cooled, higher pressures and warmer temperatures on the valley floor would have made the Valles Marineris one of the last places on the surface capable of sustaining exposed bodies of water.

Today, however, the major factors shaping the canyon appear to be those of so-called "mass wasting"—the geological tendency of elevated areas to slump into low-lying ones over time, under the influence of gravity. This process may happen gradually, or dramatically through sudden landslips.

RIGHT An artist's impression captures the moment of separation between the Viking orbiter and its aeroshell-clad lander. Each Viking lander was released close to the "apareon" point, where the spacecraft's orbit took it farthest from Mars.

RIGHT An artist's impression captures the moment of separation between the Viking orbiter and its aeroshell-clad lander. Each Viking lander was released close to the "apareon" point, where the spacecraft's orbit took it farthest from Mars.

## Vikings in Orbit

In the mid-1970s, NASA followed up the success of Mariner 9 with Vikings 1 and 2, each consisting of an orbiting survey satellite and a lander intended to touch down on the Martian surface. The Vikings had evolved from a more ambitious series of second-generation Mars probes known, somewhat confusingly, as Voyagers. Planned from 1966, these huge 12-ton (11-tonne) space missions would have been launched in 1973 using the huge Saturn V rockets already developed for the Apollo manned lunar program, but as the scope of the Apollo program was curtailed from the late 1960s onward, they were left without a suitable launch vehicle.

In the wake of this disappointment, engineers at NASA's Jet Propulsion Laboratory came up with a scheme for delivering a scaled-down mission at considerably lower cost, and so the Viking program was born (with the name Voyager recycled for the famous pair of probes to the outer planets).

Each mission consisted of an orbiter (modified from the basic design of Mariner 9, but equipped with significantly more advanced instruments) and a lander that was much smaller than those planned for the Mars Voyagers. A total fueled weight of 7,776 lb (3,527kg) allowed them to be launched using a Titan-III-E/Centaur rocket combination, much smaller and cheaper than the Saturn V.

Nevertheless, wasted development costs from the earlier plans still meant that the overall cost of Viking reached more than a billion dollars, and, once inflation is taken into account, it retains the record for the most expensive unmanned space program ever undertaken.

Viking 1 was launched on August 20, 1975, and arrived in Martian orbit on June 19, 1976, while Viking 2 was launched on September 9, 1975, and had a slightly longer journey, reaching orbit on August 7, 1976. Each spacecraft spent roughly a month on a preliminary survey of possible landing sites before releasing its lander toward the surface. Freed from the burden of their payloads, the orbiters then began their primary mission of mapping the Martian surface at far higher resolution than that achieved by Mariner 9.

Each spacecraft carried just one main scientific instrument, the Viking Visual Imaging Subsystem. This was a pair of television cameras, each attached to a small telescope and mounted on a scanning platform that could be controlled independently from the orbiter's inclination. A set of six filters could be placed in front of the camera, allowing mission scientists back on Earth to reconstruct full-color images from sets of monochrome ones.

At a typical orbiter altitude of about 1,000 miles (1,600 km), each camera could capture a region of the Martian surface roughly 25x28 miles (40x44 km) in a single image.

During the primary phase of their missions, the orbiters were restricted to relatively safe, high orbits, but (after a brief communications blackout as Mars entered solar conjunction in the fall of 1976), each was able to resume a more daring extended mission. Onboard rocket motors were used to lower their point of closest approach to the surface, and close encounters with the Martian moons Phobos and Deimos were also planned.

Following a leak in its propulsion system, the Viking 2 orbiter was eventually shut down in July 1978 after 706 orbits, but the Viking 1 orbiter continued to function for more than 1,400 orbits, shutting down as its batteries began to fail in August 1980. In total, the two spacecraft delivered more than 52,000 photographs, enabling scientists back on Earth to piece together high-resolution maps covering the entire planet.

RIGHT An unusual Viking view captures water-ice clouds forming close to the Valles Marineris around dawn. Orbiter missions since Viking have so far all opted for a "sun-synchronous" orbit. This has the advantage of showing areas of Mars under uniform lighting conditions, but renders views such as this, showing sunrise across a Martian landscape, unachievable. What's more, because certain weather conditions are thought to recur only at specific times of day, sun-synchronous orbits have limited our understanding of Martian meteorology.

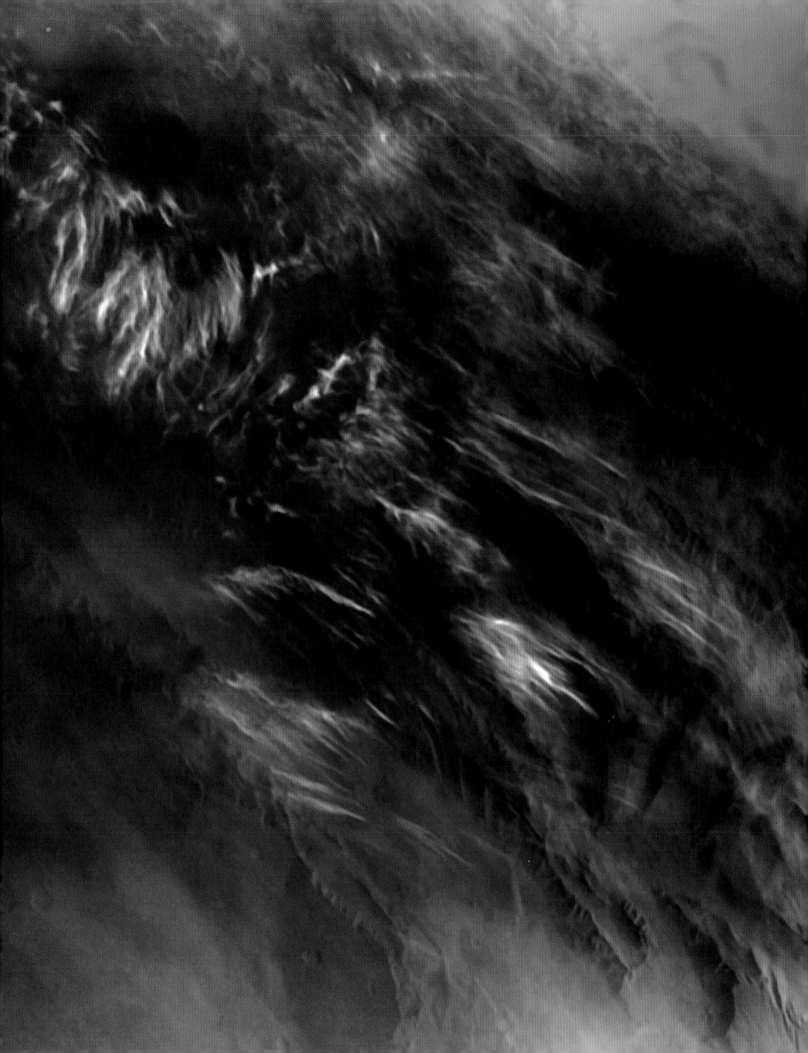

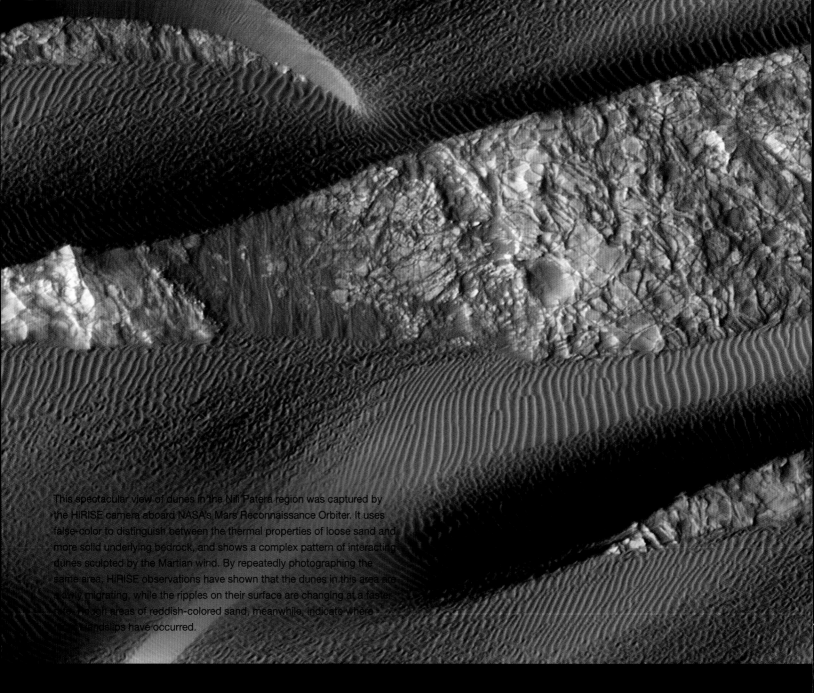

This spectacular view of dunes in the Nili Patera region was captured by the HiRISE camera aboard NASA's Mars Reconnaissance Orbiter. It uses false-color to distinguish between the thermal properties of loose sand and more solid underlying bedrock, and shows a complex pattern of interacting dunes sculpted by the Martian wind. By repeatedly photographing the same area, HiRISE observations have shown that the dunes in this area are slowly migrating, while the ripples on their surface are changing at a faster rate. Rough areas of reddish-colored sand, meanwhile, indicate where landslips have occurred.

## Martian Soil

The idea of Mars as a barren, red desert had been developing for almost a century before space probes arrived in Martian orbit, but the complexity of its shifting sands still came as a big surprise. The Red Planet's sand proved to be much finer than anyone had suspected. We now know its finer grains are an average of 3 microns (thousandths of a millimeter) in diameter, giving it a consistency similar to talcum powder. Even the larger sand grains are just tens of microns across, making them more akin to dust on Earth.

The soil and rocks get their distinctive red color from a familiar chemical compound—iron oxide, commonly found on Earth as rust. Much of the rust is thought to have originated billions of years ago when Mars was warmer and wetter and a variety of iron-rich minerals formed, but the rusting process may still continue today through the action of ultraviolet light on exposed surface rock. In general, the Martian crust seems to be much richer in iron than Earth's crust, despite the fact that

the two planets should have initially formed with similar iron content. One theory is that Earth's larger size meant that it melted more thoroughly. As a result, more of its heavy metallic constituents sank down deep into the core, a process that is known as differentiation.

The small size of Martian sand particles means they are easily picked up and redistributed by wind, even in the sparse Martian atmosphere. Indeed, it seems likely that the tiny granules probably formed as a result of wind erosion in the first place. In the absence of water to stick them together and trigger the formation of mineral clumps, there was nothing to counter the slow, ceaseless grinding down of larger particles into smaller ones.

Today, sand and dust are a major component of the upper Martian crust, or "regolith," alongside larger rocks. Darker areas are often associated with volcanic ash rich in basalt (an iron-rich mineral formed by volcanoes), while lighter ones are

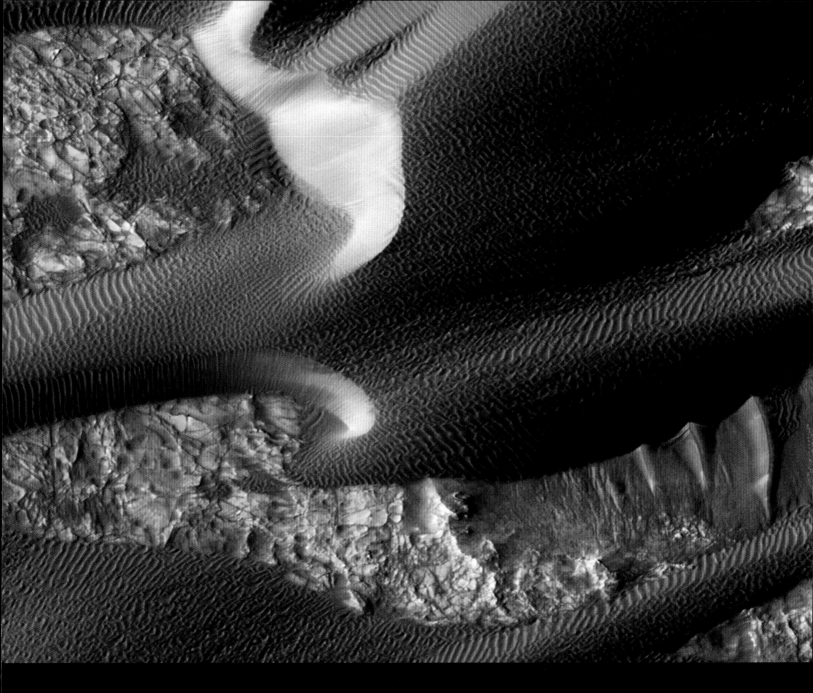

more likely to be later deposits of sand worn down from the earlier rocks. Often the layer of fine material is extremely thin, and during weather events such as global dust storms, this can be redistributed around the planet, leading to substantial changes in its appearance.

Laboratory experiments and computer models suggest that the size of dust and the density of the atmosphere together make the particles very vulnerable to "saltation," a process where they bounce along the surface on shallow trajectories like miniature cannonballs. When a particle hits the surface, it can dislodge others to create a cascadelike effect that encourages the runaway growth of massive dust storms.

The process is also very sensitive to wind speed and direction, and is largely responsible for the spectacular dune fields that cover much of the Martian landscape. Dunes on Mars vary hugely in scale and form, including many shapes seen on Earth, and some that are unique to the Red Planet.

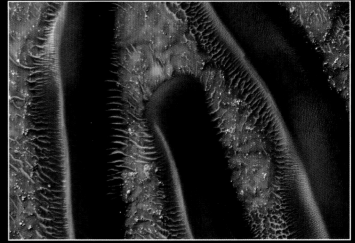

ABOVE These elongated dunes, known as millipedes, appear to have raised ridges along their sides. In fact, this is an illusion caused by an unusual lighting angle—they are really narrow gully-like channels.

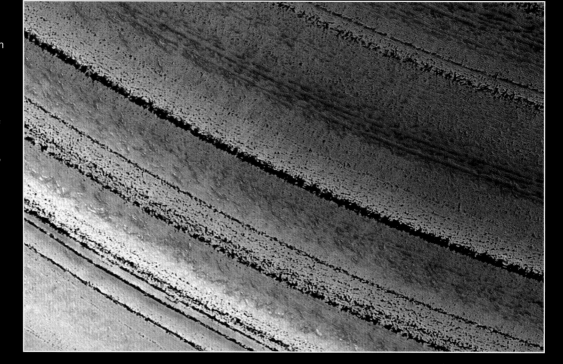

This image from Mars Reconnaissance Orbiter's HiRISE camera reveals intricate layering on the edge of the north polar cap. Like ice sheets on Earth, the permanent Martian caps preserve a record of climate conditions that stretches back many thousands of years. These results show that Nili Patera, and other regions on Mars, are areas of active sand migration and landscape erosion.

## Exploring the Ice

The arrival of space probes in long-term orbit around Mars, coupled with the first controlled landings on the surface, allowed scientists to study the planet's polar regions in detail for the first time. Even as early as the Mariner 7 flyby, the swirling appearance of the south polar cap had been obvious, and Mariner 9 revealed that the cap's surroundings had a multilayered or laminated structure. They were also scratched and pockmarked with numerous pits that were clearly not impact craters or volcanic calderas. Instead, these seemed to be signs of erosion caused by wind and far more widespread ice in the distant past.

Mariner had detected small amounts of water vapor over the Martian south pole, but the general consensus among scientists was that, since carbon dioxide was the key ingredient of the Martian atmosphere, the changing size and structure of the ice caps must be due to seasonal condensation and sublimation of $CO_2$ (sublimation is the transformation of certain substances direct from solid to gas without passing through an intermediate phase). This idea was supported by Viking lander measurements that found strong seasonal changes in overall atmospheric pressure: the density of the thin Martian air decreases by as much as 30 percent when the larger southern ice cap is at its most extensive.

It was something of a surprise, then, when the Viking orbiters detected that the residual northern ice cap, the part that persists through the summer, is made not of carbon dioxide but of water ice. This was an important hint that maybe Mars was not as dry as previously thought, and the formation of water-ice frosts around the higher-latitude Viking 2 lander site during northern winter seemed to support this idea. Scientists suspected that the southern polar cap might also have a water ice component, but despite the cap being considerably smaller (250 miles [400 km] across, compared

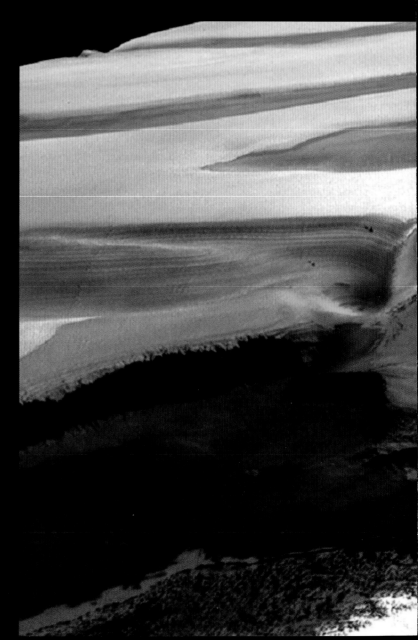

to 680 miles [1,100 km] in the northern hemisphere), it is colder, and persistent carbon dioxide ice made it impossible for the Vikings to measure its underlying composition.

The difference in temperature proved to be linked to altitude. Sitting in the middle of the enormous low-lying plain known as the Vastitas Borealis, the northern cap is considerably lower and therefore warmer than the southern one. In addition, the residual northern cap is about 1.2 miles (2 km) thick while the southern one is about 2 miles (3 km) deep, boosting its upper surface to even higher, colder altitudes.

Both permanent ice caps, meanwhile, revealed a structure of overlapping layers of ice forming a general spiral, with dark-floored canyons penetrating some way into them. The largest

of these is the northern-hemisphere Chasma Boreale, which is 60 miles (100 km) wide and 1.2 miles (2 km) deep. The spiral patterns seem to be created by the pattern of winds blowing out from the pole and controlling the deposition of ice.

More recent Martian probes have revealed many more details of Martian ice and its effect on the overall landscape. They've shown that the southern residual ice cap does indeed contain a large amount of water, but also that ice in both hemispheres extends far beyond the exposed caps, mixing with soil to create an extensive permafrost down to mid-latitudes, which can behave like pure ice in the right circumstances. The semi-permanent frosts, meanwhile, can create a variety of spectacular, though short-lived, landscape features in the polar regions as they come and go with the seasons.

**BELOW** This perspective view from Mars Express shows 1-mile (1.6-km) high cliffs and volcanic ash deposits from young volcanoes along the edge of the north polar ice cap.

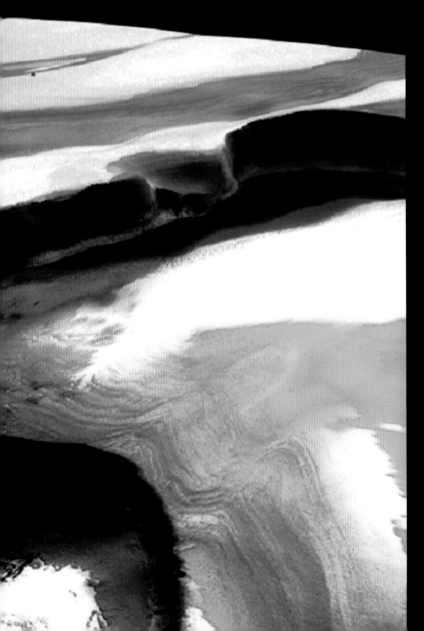

**ABOVE** This impressive Viking view shows the residual polar ice cap at the height of northern summer. The deep canyon cutting into the ice cap at upper left is known as the Chasma Boreale.

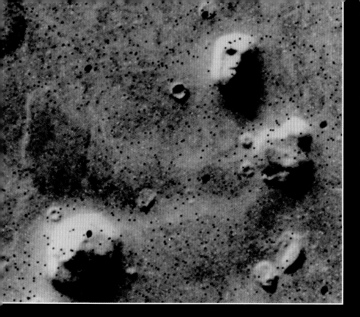

The infamous "Face on Mars" is prominent in this image from Viking Orbiter 1, taken in July 1976 during the mission's initial search for a landing site. The illusion is mostly caused by the angle of sunlight striking an eroded rock mesa.

# The Face on Mars

Perhaps the most infamous discovery made by the Viking orbiters was the so-called "Face on Mars." This misshapen rocky outcrop in the Cydonia region, when lit from just the right angle, takes on the appearance of a crudely sculpted human face.

Cydonia lies on the boundary between the southern highlands (specifically a region known as Arabia Terra) and a low-lying northern plain called Acidalia Planitia, and is littered with similar flat-topped outcrops known as *mensae* (from the Latin for "table"). Nearby lies a region of small, knoblike hills, Cydonia Colles, and a complex maze of intersecting valleys, Cydonia Labyrinthus.

"The Face" was first identified on images captured in July 1976. At first it was clearly a trick of the light, but when a second image, taken with the Sun at a different angle, showed the face even more clearly, some began to wonder whether it might be something more interesting.

Enthusiastic amateurs were soon poring over other Viking images of the area, pointing out what they maintained were unnaturally geometric features that were not simple rock formations but looked very much like the ruins of an ancient city.

# Mars Global Surveyor

Fortunately for the conspiracy theorists, who went on to weave tales of NASA cover-ups, and unfortunately for the scientists keen to debunk their fantasies, the Viking missions were followed by a long hiatus in the exploration of Mars. By the time Mars Global Surveyor arrived in orbit in 1997

A far more detailed image of Cydonia reveals the true nature of the "Face" mesa, and shows that it is one among several such outcrops in the area. Equally prominent, for example, is the apparent "skull" located to the left of the "Face."

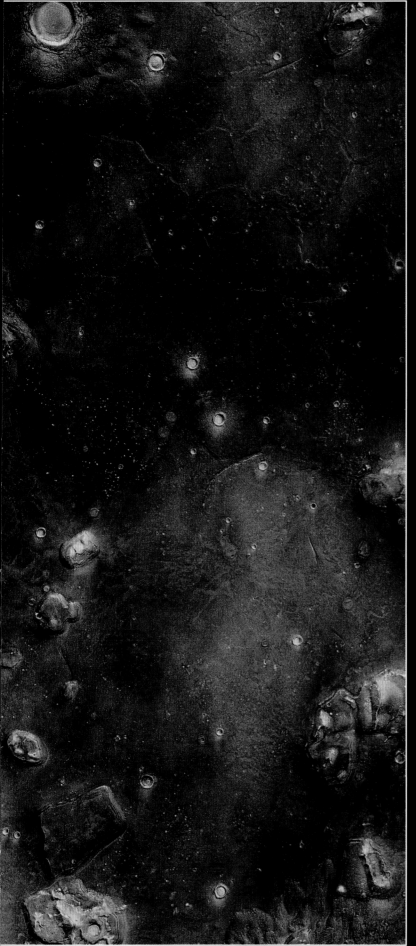

The Face had taken on a near-iconic status, and there was pressure to revisit the site as a priority.

This was not, of course, the primary purpose of the Mars Global Surveyor (MGS) mission. Unlike the Viking spacecraft, MGS had no "lander" element, but was capable of entering Mars' atmosphere, orbiting as close as 68 miles (110 km) to monitor weather, take atmospheric readings, study the topography of the planet and take a range of photographs of the surface, including high-resolution images.

MGS was originally intended to remain in orbit around Mars, transmitting data, for two Earth years (equivalent to one Martian year) but, having arrived on station in September 1997, it was still operational up to the time when contact was lost nine years later. As well as an enormous stream of scientific data, MGS transmitted more than 240,000 images back to Earth.

## Natural Rock Formation

Among many other things, the MGS images provided strong evidence, in photographs of meandering valleys possibly created by water erosion, that water had once flowed on the surface of Mars. The photographs also finally laid to rest speculation about The Face. The images revealed The Face to be a rock formation that most neutral parties, viewing the images entirely objectively, accepted was entirely natural. The mesa only takes on the appearance of a face in low-resolution images and under very particular lighting conditions. It is merely a fascinating optical illusion.

This evidence did not, however, stop conspiracists from picking over the pictures, using various techniques to "enhance" them and reveal what they claimed to be signs of artificial construction. Such theories were dealt another blow in 2006, when images from Mars Express allowed scientists to reconstruct the feature in three dimensions, confirming that it actually has a rather lopsided but decidedly natural appearance.

## Ancient Shoreline

Nevertheless, the many chaotic features of Cydonia mean it is still an interesting region for planetary scientists. Early photographs suggested that many of the mesas were surrounded by raised terraces that enhance their artificial look, and some experts suggested that they might once have been islands along the shoreline of an ancient northern ocean.

According to this theory, the terraces formed as waves wore down the edges of the islands, creating fringing shelves similar to those found on Earth. Studies of the region using satellite imagery from the most recent missions to Mars have found that the terraces are far from consistent and, in some cases, may themselves be illusions.

It seems likely that Cydonia was formed by a combination of flooding and subsidence in much the same way that other similar "chaos" regions of Mars came about. The area may have lain on the shores of a once extensive, but now vanished, Martian ocean, although the best evidence for that lies elsewhere on the planet.

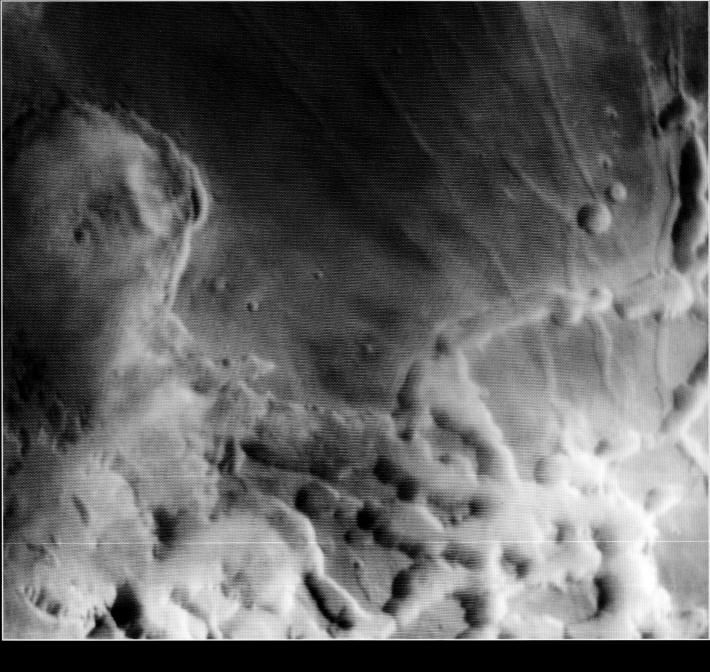

ABOVE A rare Viking image of the Martian landscape at sunrise shows part of the Noctis Labyrinthus filled with early-morning mist, probably a result of water ice frost sublimating into the air under the heat of the early-morning Sun.

# Noctis Labyrinthus

Lying at the eastern end of the huge Valles Marineris canyon system, the Noctis Labyrinthus or "Labyrinth of Night" is a spectacular region of jumbled terrain at the foot of the mighty Tharsis Rise. Initially discovered by Mariner 9, this complex network of isolated mesas and deep cracks in the Martian crust was targeted for further imaging by the Viking orbiters, and has remained a focus for study by later generations of space probes.

The Labyrinth covers an area roughly 600 miles (1,000 km) across, merging into the southwest flank of Pavonis Mons, the middle of the line of three Tharsis Montes. Viking images showed that many of the valley floors separating the flat-topped mesas were themselves surprisingly flat and, based on their numbers of accumulated impact craters, were probably of similar age to their higher surroundings.

This suggests that many of the valleys are *grabens*—a type of geological feature in which parallel faults isolate a section of landscape and allow it to slip gradually downward compared to its neighbors. Such faulting could have been triggered by the growth of the Tharsis Rise up to 2 billion years ago, and is probably linked to the opening up of the much broader Valles Marineris faults.

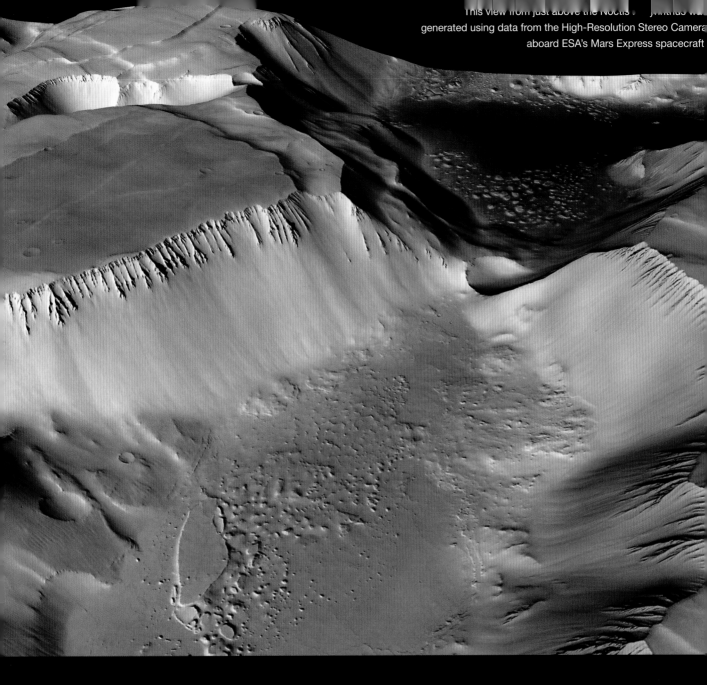

## Melting Ice

Elsewhere in the Labyrinth, the valley floors are decidedly more jumbled and chaotic. These regions may be a result of subsidence as subterranean ice melted due to volcanic activity, removing support for the overlying ground, which then collapsed. There are also many signs of subsequent landslips scattering debris from canyon walls across the valley floors.

Understandably, the Noctis Labyrinthus presented something of a headache for the scientists trying to figure out its origins, but in subsequent years the canyon network has come to be seen as a geological treasure trove.

The initial formation of grabens exposed many rocks formed in early Martian history along the steep-sided valley walls, and

later events have distributed these rocks over the floor as so-called "light-toned deposits." Analyses of these deposits in images from space probes like Mars Odyssey have suggested the presence of certain minerals, such as sulfates and clays, that are believed to have formed in the presence of water.

Even more recently, instruments on the Mars Reconnaissance Orbiter have revealed hydrated (water-bearing) minerals on the valley floors that probably formed in situ.

This would suggest that water persisted in the Labyrinth until surprisingly late in Martian history, perhaps helped by the same relatively favorable conditions that still give rise to its beautiful dawn mists today.

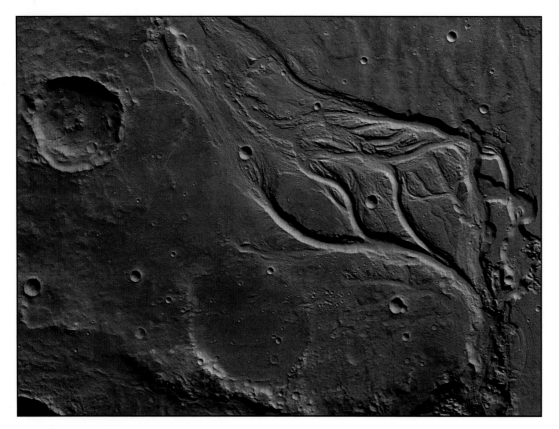

## Signs of Ancient Water?

Mariner 9 and the twin Viking orbiters began a long, slow change in the Red Planet's image. Most of the 20th century had seen the enthusiasts for a more hospitable Mars beaten into retreat by one new discovery after another, from the dismissal of the Martian canals, through the discovery of the thin, dry Martian atmosphere, to the images of barren, cratered landscapes sent back by the first Mariner flybys.

Mariner 9's discovery of volcanoes and canyons did at least reveal a more interesting geological history, but what about the prospects for past or even present life? One of the prerequisites for life as we understand it, of course, is water, but even optimists had doubts about whether Mariner could find traces of past water on the surface.

As early as 1998, astronomer Carl Sagan and others were cautioning that "we should not expect any trace of them to be visible… unless they were of greater extent than typical features on the Earth."

They need not have worried. Shortly after its arrival, Mariner 9's early photographs began to reveal narrow channels and networks of tributaries among the cratered southern highlands. Typically up to just over 1 mile (1.6 km) across, the longest of these channels run for hundreds of miles. They include Ma'adim Vallis, which follows a roughly northward track out of the highlands for around 500 miles (800 km) before terminating at Gusev Crater, and Nirgal Vallis, a 300-mile (500-km) network that begins with multiple channels that merge together into a single, deep valley.

Elsewhere, both Mariner 9 and the later Viking orbiters mapped extensive regions known as outflow channels, regions on the boundary between elevated highlands and lowland plains where the landscape appears to have been scoured clean. The most famous of these empties from the east of the Valles Marineris into the plain of Chryse Planitia. Others lead into northern lowlands such as Elysium and Amazonis Planitia, and into the huge southern-hemisphere crater of Hellas.

## Rivers and Floods

These features appear to be the unmistakeable work of water but, because the outflow channels showed no signs of tributaries, it seems that two very different mechanisms are responsible. The outflow regions probably formed in sudden, catastrophic flooding events, most likely when water was released from hidden reservoirs beneath the surface. This is supported by the fact that they often emerge from "chaos" regions where the landscape seems to have undergone dramatic subsidence, perhaps as underground water or ice was lost. The narrower valleys, meanwhile, have deep, winding features that suggest slow and steady erosion over much longer timescales—ancient rivers on the surface.

Despite this evidence, some still had doubts. While the parallels to geographical features on Earth are obvious, Mars is not Earth—conditions there are very different. Some experts suggested alternative mechanisms to form similar structures, often involving the melting and escape of carbon dioxide ice. By the time the first phase of Martian exploration came to an end with the decommissioning of Viking Orbiter 1 in 1980, the signs were there, but the jury was still out on Martian water.

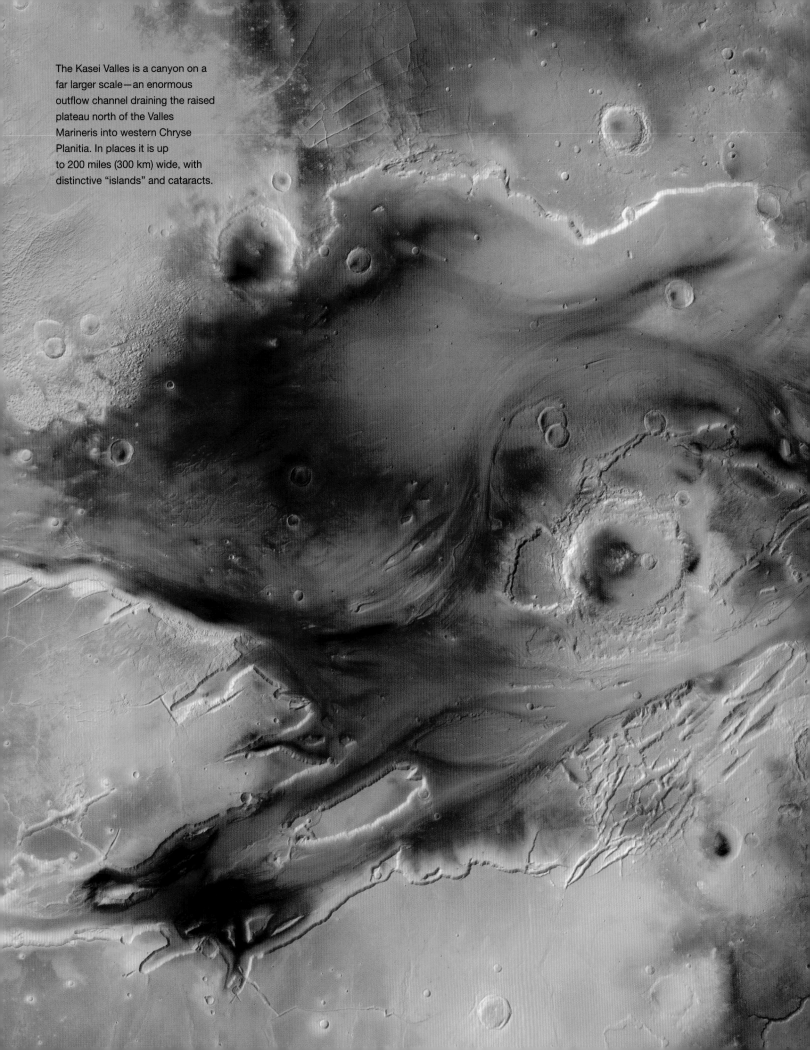

The Kasei Valles is a canyon on a far larger scale—an enormous outflow channel draining the raised plateau north of the Valles Marineris into western Chryse Planitia. In places it is up to 200 miles (300 km) wide, with distinctive "islands" and cataracts.

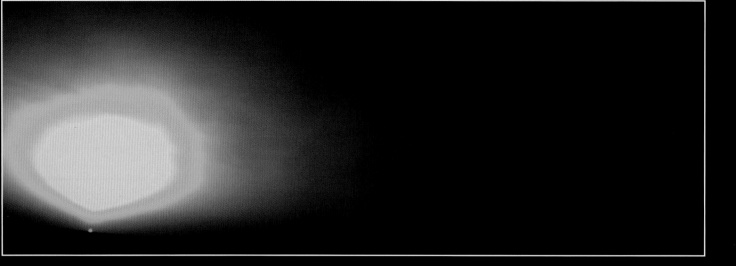

Viking Lander 2 captured this image of sunrise over Utopia Planitia on June 14, 1978, at the beginning of its 631st sol on the Martian surface. On Mars the Sun has roughly 60 percent of its diameter as seen from Earth, and provides about 45 percent of the light.

## Landings on the Surface

Along with the twin orbiters, the second key element of NASA's Viking program was the first controlled landing on Mars. The twin Viking landers were released from their respective orbiters after a month of initial reconnaissance, and landed on July 20 and September 3.

Viking Lander 1 was targeted at the intriguing outflow region on the edge of Chryse Planitia, where it might gather more evidence for suspected floods in the distant past. Viking Lander 2, meanwhile, touched down on the opposite side of the planet and some way farther north on the plain of Utopia Planitia—actually the interior of a huge, ancient impact basin.

Each Viking lander had a mass of 1,262 lb (572 kg), and made the first stage of its descent protected by an "aeroshell" with a shallow conical shape. This shield was made of ablative material designed to char under friction with the atmospheric gas, then break off, carrying away excess heat in the process.

Atmospheric re-entry alone slowed the speed of the lander's descent to about 560 mph (900 km/h), then at an altitude less than 4 miles (6 km) above the surface, parachutes opened to slow it still further. Finally, the remains of the aeroshell were released, and retrorockets on the lander's underside fired to bring the vehicle gently to a controlled landing, while leaving the soil directly beneath untouched. This was a vital aspect of the landing procedure, since studying the Martian soil was the lander's primary goal. Each lander was equipped with a robot arm for collecting samples from its immediate surroundings, and these could then be passed to instruments including an X-ray fluorescence spectrometer, a device that subjected soil samples to X-rays and studied their behavior, analyzing their composition.

## Controversial Results

Other soil samples were processed through a series of experiments aimed at detecting traces of microbial life. The results of these tests remain controversial four decades later. While most of the tests suggested the soil was sterile, the so-called "Labeled Release" experiment fed nutrients enriched with a radioactive carbon isotope to a sample of soil, and found that radioactive carbon was released back into the atmosphere above it. This behavior might be expected if microbes in the soil were processing the nutrients and releasing gas, but it was not repeated when the sample was given a second dose. Some, including the engineer who devised the "Labeled Release" experiment, argued that it had found traces of life, but other scientists put forward alternative explanations for the test results.

Vital though the soil experiments were, they were just one element of the Viking landers' scientific program. Each spacecraft was equipped with a pair of cameras to collect 360-degree panoramas, a weather station, samplers for atmospheric gases, and magnetometers and seismometers for measuring the magnetic properties of the soil and detecting possible Marsquakes. High- and low-gain antennae, meanwhile, were able to maintain direct contact with Earth, allowing the landers to carry on working long after the orbiters had ceased to function.

Viking 1's location on Chryse Planitia proved to be a sandy, rolling plain covered in rocks, probably of volcanic origin, that had later been modified by the action of ice and other forms of weathering. Utopia Planitia, in contrast, seemed to have a harder crust on which rocks are "perched," stranded after wind blew away much of the soil around their bases. By the beginning of the 1980s, we understood a lot more about Mars than we had a decade before, and data from the Viking missions would keep scientists busy for years to come—which was just as well, as it would be a long time before the next phase of Martian exploration could begin.

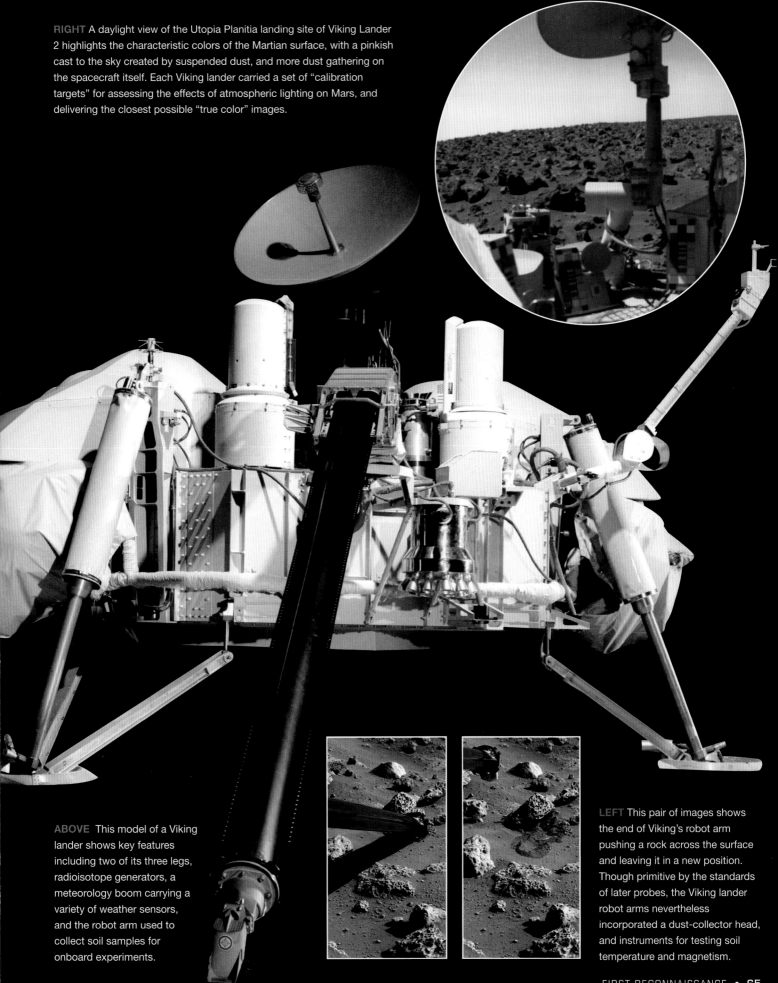

**RIGHT** A daylight view of the Utopia Planitia landing site of Viking Lander 2 highlights the characteristic colors of the Martian surface, with a pinkish cast to the sky created by suspended dust, and more dust gathering on the spacecraft itself. Each Viking lander carried a set of "calibration targets" for assessing the effects of atmospheric lighting on Mars, and delivering the closest possible "true color" images.

**ABOVE** This model of a Viking lander shows key features including two of its three legs, radioisotope generators, a meteorology boom carrying a variety of weather sensors, and the robot arm used to collect soil samples for onboard experiments.

**LEFT** This pair of images shows the end of Viking's robot arm pushing a rock across the surface and leaving it in a new position. Though primitive by the standards of later probes, the Viking lander robot arms nevertheless incorporated a dust-collector head, and instruments for testing soil temperature and magnetism.

# MAPPING THE RED PLANET

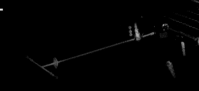

After a long pause in exploration, space agencies once again turned their attention to Mars from the late 1990s onward. Since then, a new generation of space probes has applied technology from Earth-orbiting satellites to study the Red Planet. In the process, they have made a series of astonishing discoveries that combine to make Mars seem more Earthlike than most 20th-century scientists could have dared to hope.

## First Soviet Soft Landing

Despite the achievements of the Viking missions, NASA turned its attention away from Mars in the 1980s, focusing closer to home with the ambitious space shuttle program. In the meantime, the Soviet space agency redoubled its efforts to reach the Red Planet.

Up to that point, their greatest successes had been the Mars 3 and Mars 5 probes. Mars 3 made a successful soft landing in 1971, but communication with the lander was lost

within 15 seconds, before any meaningful data could be sent. Mars 5, an orbiter that reached Mars in February 1974, began returning data, only to fall silent after nine days. It is believed that micrometeorite damage compromised the craft's pressurized instrument compartment, and the equipment was not designed to work in an unpressurized environment.

The Soviets next planned a pair of ambitious missions known as Phobos 1 and 2, which would not only survey Mars from orbit, but also drop probes onto its largest moon. Launched on July 7, 1988, a technical glitch caused Phobos 1 to lose contact with its controllers on Earth while still en route to Mars.

Phobos 2 was launched five days after its sister craft and spent almost seven months on its journey to Mars. When it arrived in January 1989, it began sending back data, but it also lost contact shortly before its planned close encounter with Phobos.

## Disastrous Fuel Leak

NASA's next mission, meanwhile, was Mars Observer, an orbiter scheduled to arrive at Mars in 1992. Fitted out with the kind of "remote sensing" instruments that had been revolutionizing satellite studies of Earth, Mars Observer also fell victim to what some were now calling "The Curse of Mars."

Having enjoyed a successful launch from Cape Canaveral in Florida on September 25, 1992, Mars Observer was crippled by what may have been a leak of pressurized fuel, which sent the craft into a spin and also damaged electrical circuits. Mars Observer had been just about to enter orbit around Mars in August 1993 when all communication with Earth was lost.

## Rocket Failure

Three years later, Russia's ambitious Mars '96 probe re-used

ABOVE A model of Russia's failed Mars '96 probe. This mission consisted of both an orbiter and landers to study and penetrate the Martian soil.

Unfortunately, Mars '96 suffered a catastrophic failure after launch when the rocket engine intended to take it out of Earth orbit and send it on its way to Mars failed to fire. The spacecraft broke up on re-entry, although some elements are believed to have crashed into the Pacific Ocean and the mountains near Chile's border with Bolivia.

## NASA's New Strategy

For NASA, the failure of Mars Observer had been one among many setbacks, others including the technical faults that initially crippled both the Hubble Space Telescope and the Galileo Jupiter probe. As a result, NASA rebuilt its unmanned space program around a different, "faster, better, cheaper" philosophy, replacing ambitious and expensive all-purpose spacecraft with focused missions that carried a more limited range of scientific instruments.

This led to the foundation of the Mars Exploration Program, a series of missions aimed at paving the way for manned Martian landings and answering key questions about the planet's potential for habitability, its past climate, and the possibility of life. Eventually, this policy would bear fruit in spectacular style, both on Mars and elsewhere in the solar system, but there were still to be failures along the way.

RIGHT This artist's impression shows a Deep Space 2 penetrator burrowing into the Martian rock. Consisting of two such probes, Deep Space 2 was carried to Mars on board the Mars Polar Lander. After reaching Mars on December 3, 1999, the probes were released, but contact was lost after they hit the surface.

**LEFT** This artist's impression shows Mars Global Surveyor in orbit above Olympus Mons. The space probe's fuel-efficient aerobraking relied on the resistance generated by its winglike solar panels in the upper atmosphere, but unexpected movement during the first aerobraking maneuvers revealed that one panel had not latched properly into place. As a result, a longer but less intensive braking program was adopted to protect the vulnerable panel.

## Return to Orbit

The first successful probe to reach Mars in more than two decades was Mars Pathfinder, a lander with a small robot rover that paved the way for later generations of automated explorers. This lightweight mission had overtaken another spacecraft en route to Mars—the larger Mars Global Surveyor. Launched almost a full month before Pathfinder, on November 7, 1996, Global Surveyor arrived more than two months later, entering orbit on September 12, 1997. While Pathfinder's mission on the surface lasted only a month, Surveyor would send data back to Earth for nine years.

Mars Global Surveyor inherited many of the key instruments lost in the failure of Mars Observer, including the main Mars Orbiter Camera (MOC) and a Thermal Emission Spectrometer capable of mapping large-scale mineral distribution based on infrared (heat-radiation) emissions from the surface.

Months of aerobraking (a fuel-saving maneuver that adjusted the spacecraft's orbit by skimming the planet's atmosphere) brought Surveyor into a "Sun-synchronous" orbit that looped over each of the planet's poles at an average altitude of 235 miles (378 km). This allowed the cameras to map the

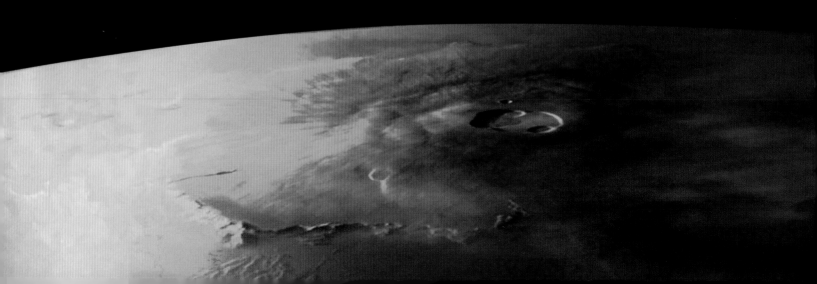

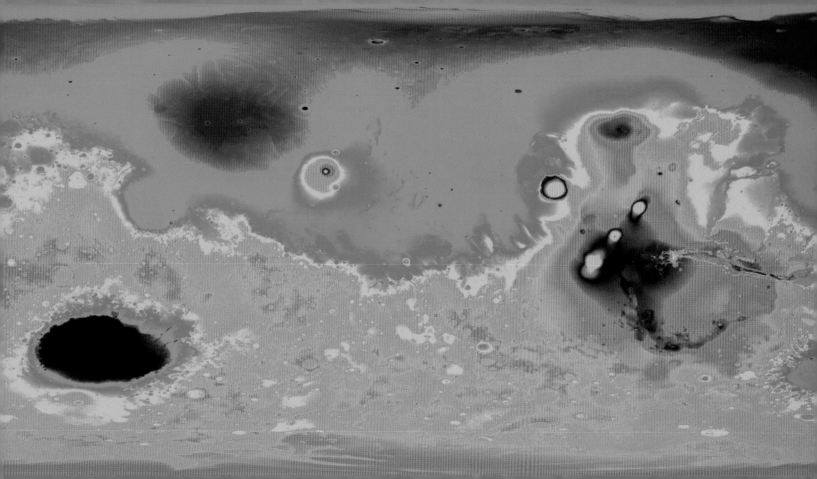

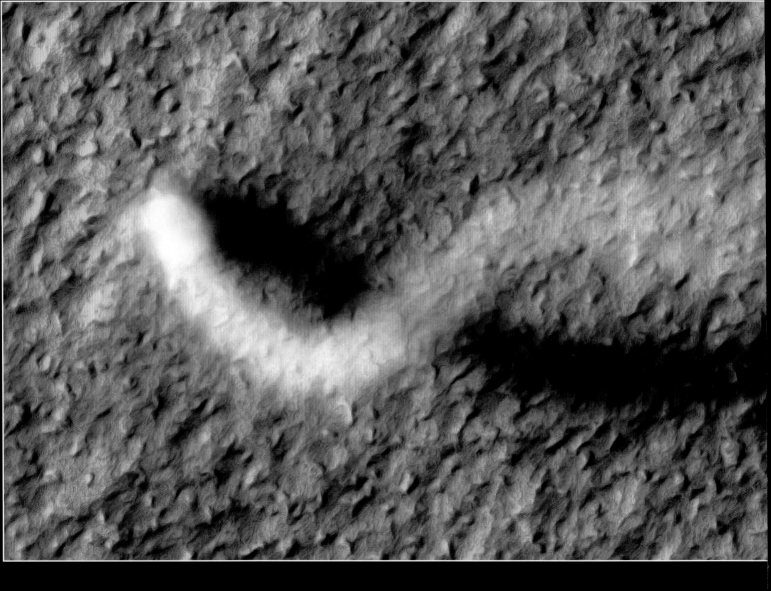

## Dust Devils

Among Surveyor's many discoveries were the spectacular but short-lived and powerful whirlwinds known as dust devils. Their existence was initially inferred shortly after Surveyor arrived in orbit, from images of dust plains that appeared to be covered in curious doodlelike markings. These swirls are created when the passage of a dust devil clears the uppermost layer of light-colored dust from the surface and exposes darker materials beneath. Similar markings have since been identified in Viking images. Nevertheless it was to be another two years before Surveyor got lucky and caught a dust devil in the act of drawing its Martian graffiti.

The Red Planet's whirlwinds are thought to have similar origins to Earth's own dust devils, generated when warm air is trapped near the ground by the pressure of cooler air above it, and then suddenly released by a shift in the overlying layer. As the warmer air funnels upward, it cools, encouraging more air to follow it. The result is a rising column of rotating air, the base of which draws in warm air from its surroundings. As the whirlwind concentrates ever more rotating air into a narrowing column, it naturally spins with increasing speed. Strong surrounding winds dislodge material from the ground, and the intense low-pressure zone at the center draws it upward.

The major difference between Martian dust devils and their terrestrial equivalents is their scale. On Earth, these features are typically several feet across and up to hundreds of feet high, while devils on Mars can be ten times that size, rising high into the atmosphere. Wind speeds around them can reach more than 60 mph (100 km/h), but the low atmospheric pressure robs them of destructive power.

In fact, encounters with dust devils have proved positively beneficial for rovers on the Martian surface, scrubbing their solar panels free of the dust that inevitably accumulates over time, boosting their power supplies, and dislodging particles trapped in their delicate mechanisms.

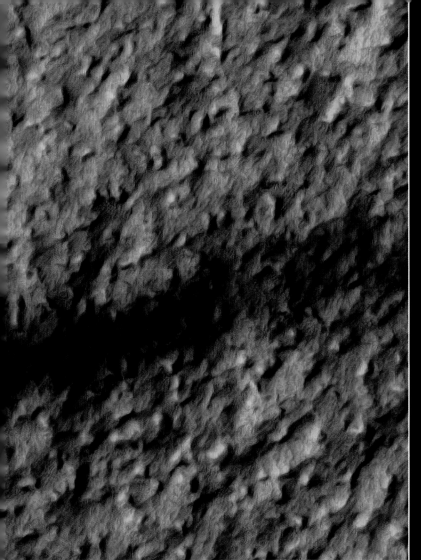

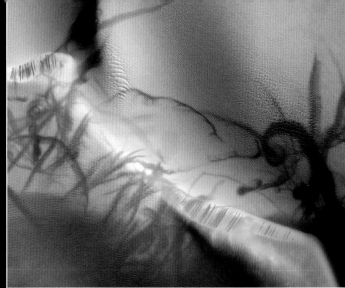

ABOVE Sand dunes in the Nili Fossae region are covered in dark scribbles and loops—a result of whirlwinds scouring away the light-colored surface dust to expose darker soils beneath.

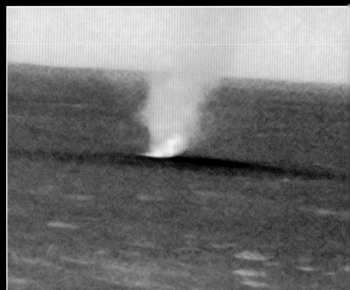

ABOVE The Mars Exploration Rover, Spirit, caught sight of this dust devil in Gusev Crater shortly after it benefited from the first of many "cleaning events" in which similar whirlwinds cleared accumulated dust from its solar panels.

## Not So Different After All?

Another important discovery, courtesy of both Surveyor and *2001 Mars Odyssey*, is that the Red Planet is far more extensively cratered than previously suspected. After identifying as many impact craters as they could in Viking images, planetary scientists concluded that the southern and northern hemispheres of Mars have radically different histories, and that the low-lying northern plains never suffered the same level of bombardment from space as the southern highlands. One popular theory was that the northern half of the planet remained molten for longer, perhaps as a result of one or more huge meteorite impacts that caused it to melt and re-form.

Detailed images from the new orbiters suggested that these old theories were mistaken. The northern hemisphere was probably just as heavily cratered, but had subsequently been blanketed with a thick layer of volcanic ash and lava in many places. Today, it seems, wind erosion is gradually wearing down the overlying material, revealing the ghostly outlines of older craters beneath. Even though the "hemispheric dichotomy" between the northern and southern halves is real, it seems to be rather different in nature from what was

previously suspected. The new evidence suggests that, even if the difference between the hemispheres was caused by huge ancient impacts, these happened very early in Martian history. The entire planet was certainly solid by the time of the Late Heavy Bombardment about 4 billion years ago. On the other hand, it seems to raise new questions about why such thick layers of volcanic material built up only in the northern hemisphere—can this be entirely due to its radically lower average altitude?

## Crater Discoveries

As well as mapping the large-scale distribution of impact craters, Surveyor and other recent satellites have uncovered the features of individual craters in much greater detail, often revealing secrets from the planet's ancient history. The ejecta blankets around many craters (fields of debris thrown out during the initial meteorite impact) often show lobed "splash" patterns that point to the presence of liquid during their formation. This may have come from ancient standing water or the sudden melting of subsurface ice.

Other craters show distinctive rock layers. Sometimes these are found in their walls, hinting that the entire surrounding landscape may be composed of layered sedimentary rocks. Elsewhere they are found on the crater floors themselves, suggesting that they were laid down after the crater formed, and have subsequently

been eroded once again. The two most likely mechanisms for creating sedimentary rocks are the repeated laying down of wind-blown dust and ash, or their deposition as mud in standing bodies of liquid.

## Shaping the Landscape

By maintaining a presence in Martian orbit over almost two decades, Mars Global Surveyor and its successors have also been able to identify short-term changes in the Martian landscape. Some of these changes have been found by targeted repeat viewing of the same area but, occasionally, changes are noticed simply because some areas of the Red Planet have now been imaged several times over the years.

Because both Surveyor and Mars Reconnaissance Orbiter follow Sun-synchronous orbits, all images for a given location on the surface are always photographed at the same time of day, and (allowing for seasonal variations in the altitude of the Sun) under similar lighting conditions.

This makes changes on the Martian surface much easier to spot, and features that have so far been found include recent landslips and newly formed impact craters.

**BELOW** Mars Reconnaissance Orbiter captured the dramatic splash of a newly formed impact crater in late 2013. The crater itself is around 100 ft (30 m) in diameter, but it is surrounded by bright rays of ejected debris extending as far as 10 miles (16 km).

## The Gully Conundrum

Features known as gullies may prove to be the most important of all Mars Global Surveyor's discoveries. These are narrow channels that appear to run downhill on steep slopes such as crater walls, and which often terminate in a fan of deposited sediments. In June 2000, NASA released images that showed several dozen of these channels in the Gorgonum Chaos region, running down dunelike slopes from the edge of elevated mesas to a lower valley floor. These gullies bear an irresistible resemblance to similar features on Earth that form when water runs downhill, carrying away loose, eroded soil.

Based on the fresh, undegraded appearance of the sediment fans and their lack of overlying impact craters, researchers concluded that the gullies are a few million years old at most. Many have seized on the discovery as vital evidence for water

running on the surface of Mars in the recent past. But as more gullies have come to light, so have significant differences between Martian and terrestrial gullies.

Most interesting is the fact that they tend to begin some way down the slope, rather than right at the top of it. In the case of Gorgonum Chaos, for instance, they appear to emerge from a layer about 300 feet (100 m) beneath the neighboring mesa. Often the top of a gully is marked by a series of small rivulets that emerge from an alcove before coming together to form the main channel.

This sort of structure makes sense if the gullies are formed by liquid escaping or seeping from an underground layer, rather than (as on Earth) from surface water encountering a slope. Despite the low temperatures, the feeble atmospheric

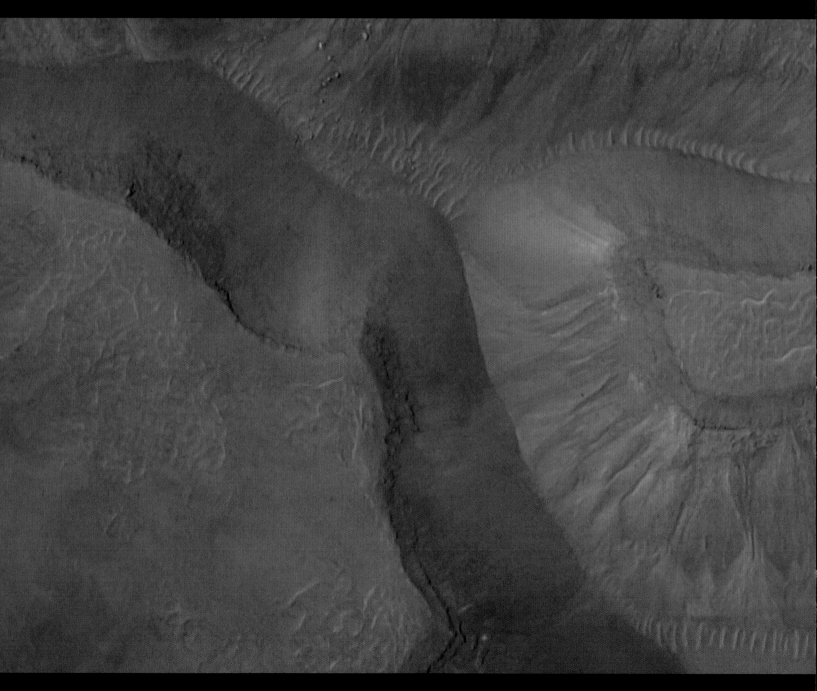

pressure on Mars means that water cannot stand or flow far on the surface without boiling away into vapor. Any liquid water is more likely to exist as subterranean water tables and it may only escape when these underground layers are cut by features such as valleys or craters.

## Carbon Dioxide Frost

Yet liquid water is not the only proposed explanation for the gullies. Another early suggestion was that these features might be created by subterranean liquid carbon dioxide kept in an unnatural fluid state by the pressure of its surroundings and violently evaporating as it reached the surface. Before too long, however, most scientists had dismissed the idea. The gullies show little sign of violent activity and the narrow, sinuous shape of their main channels suggest they were carved out over relatively long periods by a persistent force.

The Martian gullies have been subject to intense scrutiny from both Surveyor and Mars Reconnaissance Orbiter. Repeated imaging has confirmed that some gullies are actually growing and changing from year to year, but also suggests that water may not play the key role once expected.

Changes to gullies, it seems, happen mostly around Martian winter, when any water ice should be deep frozen, and the most popular current theory to explain gully formation is that they arise when winter carbon dioxide frosts sublimate back into gas, creating a lubricating effect that produces small landslips. Once a gully begins to form, the additional shade provided by its profile may encourage the build-up of frost in subsequent winters, causing growth to continue. Despite this, it seems likely that we still don't fully understand the story of the Martian gullies.

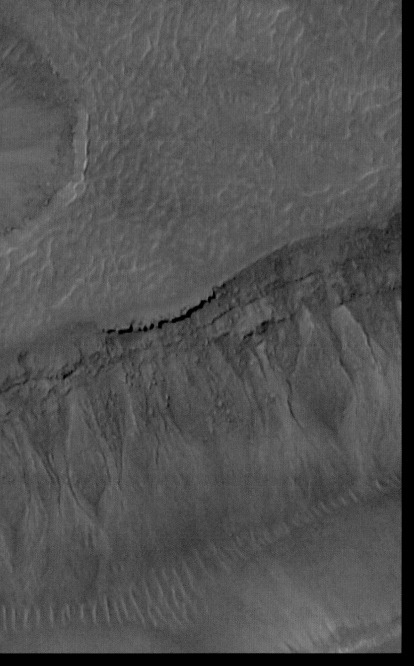

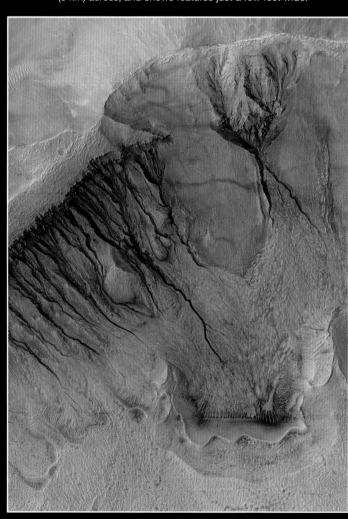

**LEFT** Mars Global Surveyor was the first spacecraft to spot the distinctive gullies within Gorgonum Chaos. This image, captured in 2000, covers an area of the surface roughly 2 miles (3 km) across, and shows features just a few feet wide.

**ABOVE** This Global Surveyor image shows an oblique view of the wall of Newton Crater, in the Sirenum Terra region of the southern highlands. Here, gullies are probably caused by the effects of melting snow.

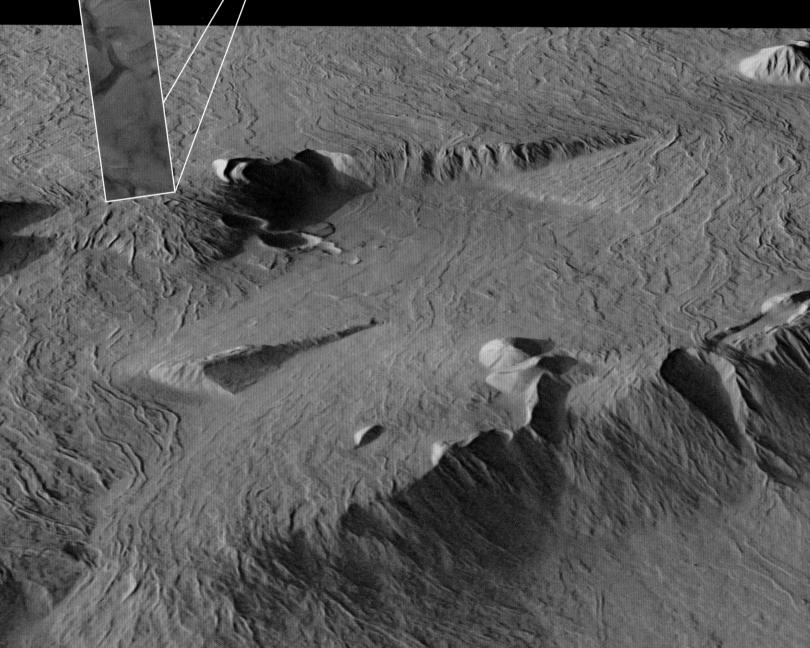

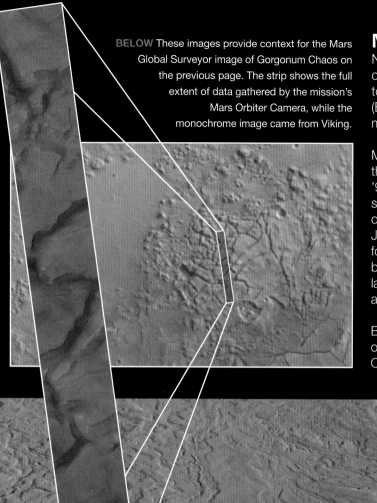

BELOW These images provide context for the Mars Global Surveyor image of Gorgonum Chaos on the previous page. The strip shows the full extent of data gathered by the mission's Mars Orbiter Camera, while the monochrome image came from Viking.

## Mars in Three Dimensions

NASA used the 2003 launch window (the closest opposition of modern times) to launch its Mars Exploration Rovers toward the Red Planet, but the European Space Agency (ESA) also took this opportunity to launch a Mars orbiter mission of its own.

Mars Express carried new versions of a number of instruments that ESA had originally developed for Russia's ill-fated Mars '96 mission, allowing it to be developed on a comparatively short timescale and tight budget in time for the 2003 opposition. Launched using a Russian Soyuz rocket on June 2, 2003, it entered Martian orbit on Christmas Day, following one of the shortest routes possible. Six days before arrival, it had released a small British-designed lander called Beagle 2, which entered the atmosphere just as Mars Express arrived in orbit.

Beagle 2 was equipped with experiments to search for traces of Martian life, but no signals were received following its Christmas Day descent. Eleven years later, NASA's Mars

Reconnaissance Orbiter located Beagle 2 on the surface, and sent back pictures showing that the spacecraft had made a safe landfall but failed to deploy two of its solar panels.

The Mars Express orbiter more than compensated for the Beagle 2 disappointment, returning stunning results over more than a decade. The most visually impressive of these came from its High Resolution Stereo Camera (HRSC), a device that has mapped features across the entire surface down to 30 ft (10 m) across, and some specific targets at five times that resolution. As its name suggests, it used a stereo camera, capable of imaging targets from two slightly different angles, allowing the construction of three-dimensional images and animations of the Martian surface.

Images from HRSC have transformed our understanding of many Martian features, revealing the true shape of "The Face" in Cydonia, showing details in the summit caldera of Olympus Mons, and mapping the complex canyons of the Valles Marineris. HRSC also played an important role in identifying landing sites for later missions.

## The Northern Ocean

Other Mars Express instruments included a suite of spectrometers for mapping the composition of the landscape, and monitoring the chemistry, temperature, and pressure of the atmosphere. In 2005, these confirmed the presence of various hydrated sulfates and silicates in the upper layers of the Martian crust, an important indication that Martian rocks have been modified by water in the past.

The spacecraft also carried a radar sounding instrument called MARSIS for mapping the crust below the surface. The strength with which radio waves are reflected back from up to just over 1 mile (1.6 km) deep is governed by a property called their "dielectric constant," which can indicate the density of the rocks and the presence of other materials such as ice.

After initial problems deploying one of the experiment's radar antennae were overcome, MARSIS has made some key discoveries, including evidence supporting the past existence of an enormous ocean that once covered much of the planet's northern hemisphere.

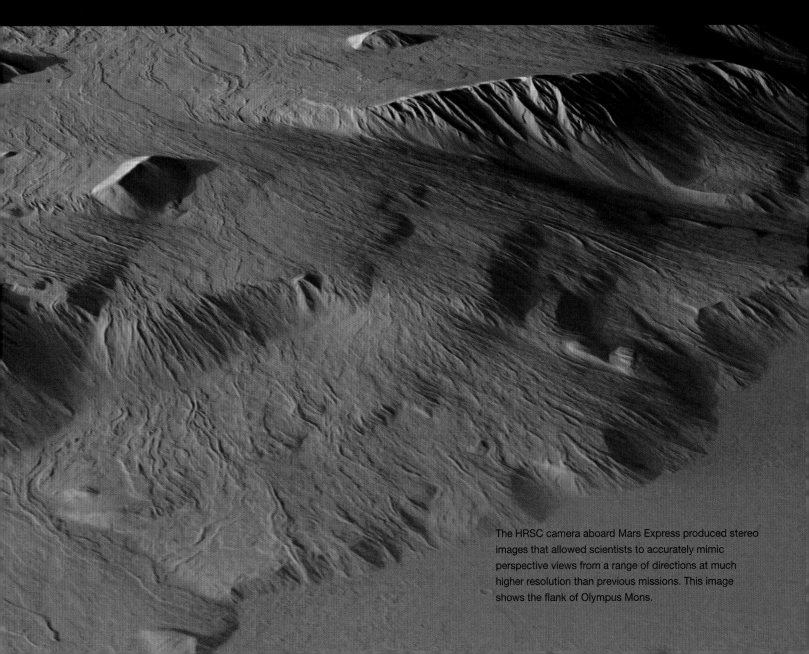

The HRSC camera aboard Mars Express produced stereo images that allowed scientists to accurately mimic perspective views from a range of directions at much higher resolution than previous missions. This image shows the flank of Olympus Mons.

## Martian Mineralogy

Following the failures of Mars Climate Orbiter, Mars Polar Lander, and Japan's Nozomi mission in 1999, the new millennium brought better luck, as NASA's *2001 Mars Odyssey* reached orbit successfully on October 24 that year. Originally named the Mars Surveyor 2001 Orbiter, it was initially planned as one half of an orbiter/lander pair, but the companion mission was put on hold after the 1999 failures (later resurrected as the 2008 Phoenix mission), and the orbiter was renamed in honor of Arthur C. Clarke's seminal science-fiction movie, *2001: A Space Odyssey.*

*Odyssey's* principal goal was to survey the composition of the Martian crust in order to find evidence of ice and water below the surface. To achieve this, it used two main instruments— the Gamma-Ray Spectrometer (GRS), and the Thermal Emission Imaging System (THEMIS).

BELOW This image from Mars Reconnaissance Orbiter shows data from the probe's Thermal Emission Spectrometer instrument. Red tones identify a concentration of hematite ($Fe_2O_3$) on the surface.

GRS was based on a Russian design initially developed for the failed Mars Observer mission of 1993 and was built around a crystal of pure germanium weighing 2 lb 10 oz (1.2 kg). The spectrometer worked by detecting electric current flowing through the crystal when it was struck by high-energy gamma radiation from the Martian surface. Gamma rays are emitted from atoms when they are hit by cosmic rays—energetic, fast-moving particles generated by the Sun and other sources in more distant space. The gamma-ray emissions from atoms of specific elements are unique, allowing the GRS instrument to help identify the distribution of key elements. The concept was proved in spectacular style when, in 2002, scientists released a GRS map of hydrogen distribution in the Martian soil, providing some of the most important evidence so far for widespread water or ice on Mars.

THEMIS, meanwhile, borrows techniques from remote-sensing instruments on Earth-based satellites, specifically "multispectral imaging," where the same area is imaged simultaneously at a variety of wavelengths. The THEMIS instrument combines two different cameras.

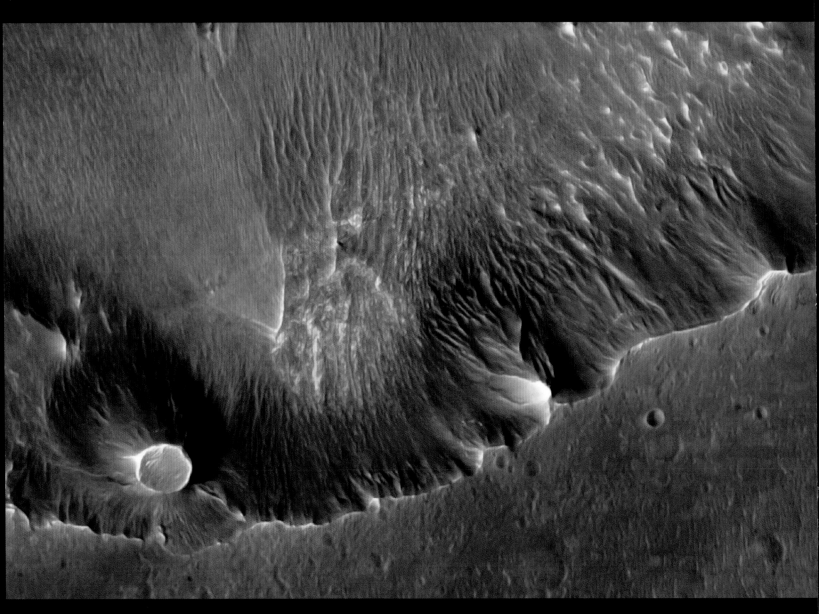

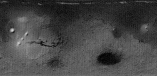

| Hydrated mineral sites | Olivine | Pyroxene | Ferric oxide | Dust |

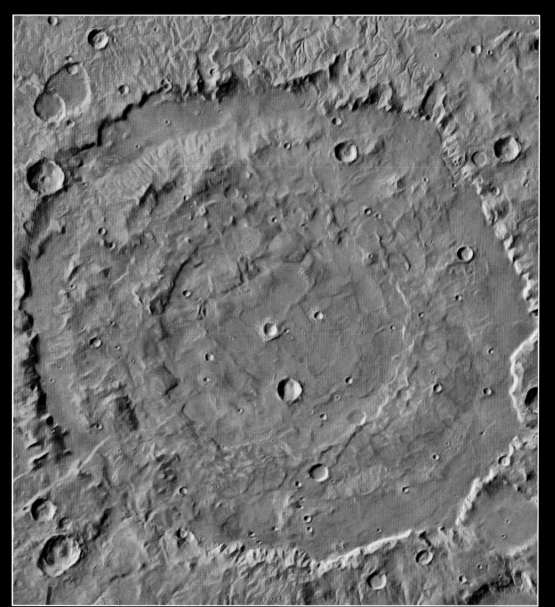

**ABOVE** A sequence of maps compiled using the OMEGA mineral mapper aboard Mars Express shows concentrations of hydrated minerals, drier olivine and pyroxene, ferric oxide, and dust. Such global maps help to identify different major rock units on the Martian surface, and major phases in the Red Planet's history.

**LEFT** When Mars Reconnaissance Orbiter mapped the 283-mile (456-km) impact basin Huygens, it found more than expected, identifying a deeply buried layer of carbonate exposed in the smaller crater at the ten o'clock position. Such deposits, if widespread, could be linked to the disappearance of most of Mars' carbon-dioxide-rich atmosphere.

ABOVE The Mars Reconnaissance Orbiter was equipped with the largest telescope ever carried beyond Earth's orbit, and a high-gain antenna capable of sending record-breaking amounts of data back to Earth—more than all previous interplanetary missions combined.

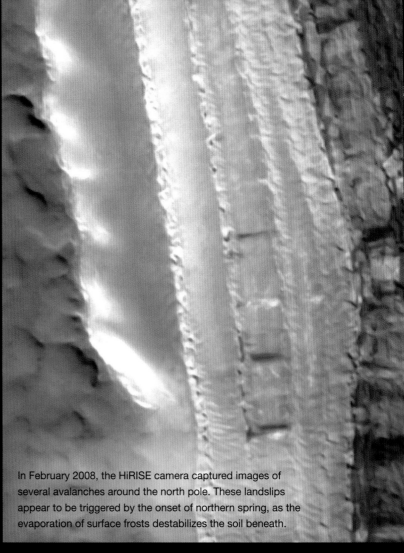

In February 2008, the HiRISE camera captured images of several avalanches around the north pole. These landslips appear to be triggered by the onset of northern spring, as the evaporation of surface frosts destabilizes the soil beneath.

## The Next Step

Launched in August 2005, Mars Global Surveyor's successor, the Mars Reconnaissance Orbiter (MRO), was to be the most ambitious Mars orbiter so far. In a throwback to the complexity of the failed Mars Observer mission, the spacecraft carried an array of sophisticated instruments to study the Red Planet in a variety of different ways. MRO reached its intended orbit without any significant problems in March 2006, and soon began sending back a wealth of data. MRO's primary instruments included three cameras, two spectrometers and a ground-penetrating radar. The most advanced of the cameras is the High-Resolution Imaging Science Experiment (HiRISE), an electronic camera attached to a 19.7-inch (0.5-m) telescope (the largest ever carried beyond Earth orbit).

HiRISE images are long, narrow strips of the surface in three different wavelengths, blue-green light, red, and near-infrared. These readings can be combined to produce true-color images or reveal other properties of the surface materials. It can identify features as small as 1 foot (0.3 m) across, similar to the capabilities of the current generation of Earth remote-sensing satellites. The Context Camera (CTX), meanwhile, captures broader monochrome images of the areas studied by other instruments. The third camera, MARCI, has a much wider field of view and is used to monitor the Martian weather.

The two spectrometers are known as CRISM (the Compact Reconnaissance Imaging Spectrometer for Mars) and MCS (the Mars Climate Sounder). CRISM splits visible light and near-infrared radiation from the surface into 544 separate channels in order to identify the characteristic properties of different minerals. The operating principle is similar to that of the TES and THEMIS instruments on Mars Global Surveyor and 2001 Mars Odyssey respectively, but CRISM offers a further step forward in detail, capturing features down to a resolution of 60 feet (18 m). MCS, meanwhile, points toward the Martian horizon as seen from orbit, studying emissions from narrow strips of the atmosphere, mostly in the far infrared, to measure temperature and pressure characteristics.

The orbiter's radar instrument is SHARAD (Shallow Subsurface Radar), a ground-penetrating radar designed by the same Italian Space Agency team that built the MARSIS instrument on Mars Express. The two are intended to complement each other. SHARAD can resolve smaller layers as thin as 23 feet (7 m) over areas hundreds of feet across, but can only penetrate to about 3,300 feet (1 km) beneath the surface. In 2009, it was able to confirm that the layers of ice in the northern polar ice cap that have accumulated due to seasonal freezing are equivalent to about one third the volume of the Greenland ice sheet. MRO made observations in conjunction with Mars Global Surveyor for several months before the older satellite lost contact. Since then, it has kept watch over the Red Planet, acting as a communications relay for current landers and assessing suitable locations for future missions.

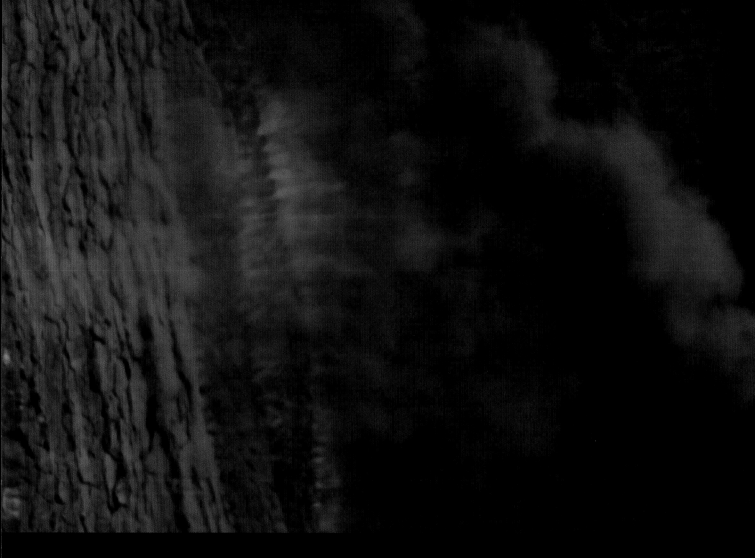

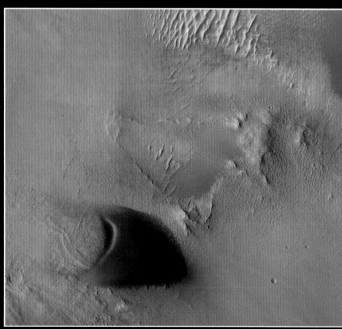

**ABOVE** MRO's powerful telescope allows it to see Mars rovers and other manmade objects on the planet's surface—here, the Curiosity rover (circled) is seen close to the center of Gale Crater's Pahrum Hills.

**ABOVE** HiRISE's imaging capabilities allow scientists to produce enhanced-color views of the surface, highlighting subtle differences such as those between fine red dust and darker, coarser sands, as seen here.

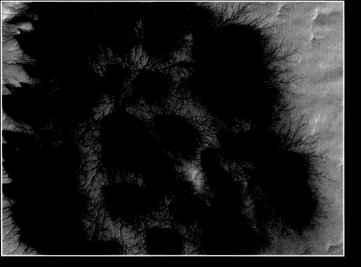

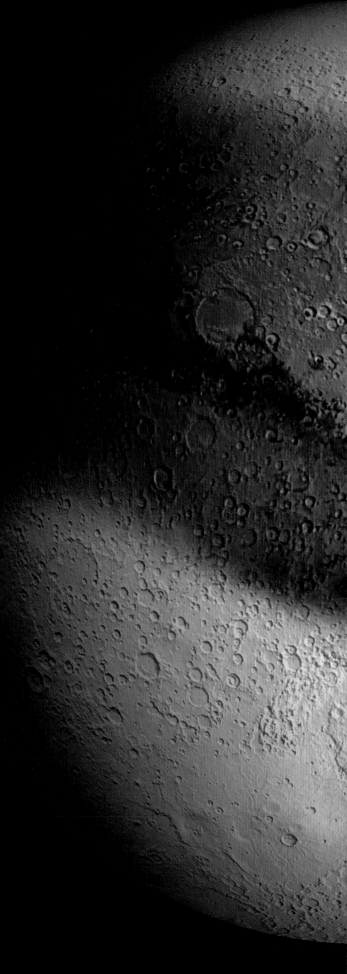

ABOVE These curious starburst or spider-patterns in the
southern polar cap form when thick layers of carbon dioxide
frost, formed during the winter, begin to evaporate at the onset
of spring. Small concentrations of surface dust are thought to
increase the heating effect of the Sun, warming the ice until it
melts just beneath the surface. Eventually, the melted material
escapes through narrow cracks as jets of dust-laden gas.
Over time, the cracks extend to form radial networks.

## Martian Climate Change

One key discovery from ongoing satellite observations
has been the fact that Mars seems to be going through
a period of long-term global warming.

The planet's regular seasons see thick carbon dioxide
frosts alternately condensing and sublimating at
opposite poles, but studies of the south pole suggest
that the underlying ice is also undergoing a decline.

Images from Mars Global Surveyor taken several years
apart showed the growth of shallow, linked pits, known as
"Swiss cheese" formations, in the residual carbon dioxide
cap previously assumed to be a permanent feature.

It seems that not only is the southern winter currently
unsuitable for the formation of ice that will persist through
the next summer, but that the existing carbon dioxide ice
is also suffering a steady year-on-year erosion.

So what could be the cause of such long-term changes on
Mars? They are most likely linked to changes in the Red
Planet's orbit and orientation known as Milankovitch Cycles.

Discovered by Serbian astronomer Milutin Milankovitch a
century ago, such cycles affect all the planets, and on Earth
they are thought to play an important role in creating cycles
of cold glacials and warm interglacials as seen during the
most recent Ice Age.

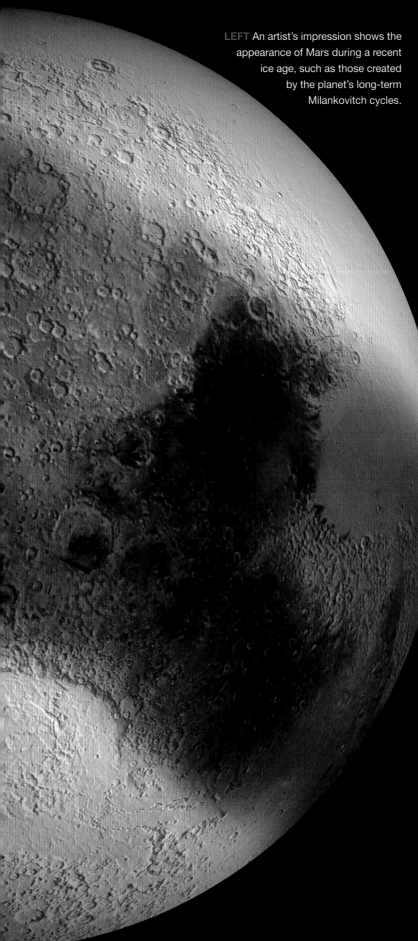

An artist's impression shows the appearance of Mars during a recent ice age, such as those created by the planet's long-term Milankovitch cycles.

On Mars, there are three distinct cycles. One is a roughly 100,000-year variation in the shape of the Martian orbit around the Sun. While today this orbit is notably elliptical, at one extreme it can be even more elongated, and at another it can take a more circular form. This obviously affects the overall amount of heat that Mars receives from the Sun at different points in its orbit.

## Axial Tilt

The second cycle is a variation in axial tilt. This describes the angle at which Mars' axis of rotation is "tipped over." While the planet spins as it orbits the Sun, it does not spin standing bolt upright relative to its orbit around the Sun. This means that, just like Earth, Mars will present one hemisphere closer to the Sun than the other at different stages of its orbit.

Axial tilt, therefore, affects the amount of heat and light reaching opposite hemispheres. This creates the planet's seasons and in the depths of winter at one of the poles there will be complete, permanent darkness. In the case of Mars, axial tilt varies between 15 and 35 degrees in a 124,000-year cycle, affecting the overall severity of the seasons.

## The Precession Wobble

Finally, a third cycle, known as "precession," also has an effect on the climate. Precession is a 171,000-year "wobble" in the direction that the planet's axis points in space. This affects the interplay between seasons, and Mars' changing distance from the Sun, by determining which hemisphere (if either) experiences summer around the planet's perihelion point (when it is closest to the Sun).

Today, Mars is 25 million miles (40 million km) closer to the Sun at northern midwinter than at midsummer, meaning that seasonal variations are evened out in the northern hemisphere. In contrast, the southern hemisphere experiences much warmer summers and colder winters because it is close to perihelion at midsummer and close to aphelion at midwinter.

At present, the warmer southern summer explains both the small size of the residual south polar cap, and how global dust storms typically develop in the southern hemisphere around perihelion. At other times, of course, the situation will be entirely reversed.

These three interlocking cycles create a complex pattern of cold ice ages and warm interglacials. At times, when the planet may be warmer, sublimating ice can enrich the atmosphere with both carbon dioxide and water vapor, producing a warm interglacial. But at other times, the polar ice caps may extend much farther than they do today, creating Martian ice ages.

## A Warmer Past

So, evidence of current climate change and Milankovitch cycles supports the idea that Mars has been through both warmer and colder phases throughout its history, while detailed images from orbit have revealed minerals and geographical features that are best explained by the long-term presence of water on the surface. But when and how did this change happen?

In order to support liquid water on the surface, Mars must once have had a much thicker atmosphere than it does at present. Originating from material of the protoplanetary nebula and enriched with gases belched out by Martian volcanoes, it would have been dominated by nitrogen and oxygen with carbon dioxide as a relatively small component. The atmosphere would have acted as a blanket over its surface, raising temperatures and providing enough pressure for water to remain in liquid form without rapidly evaporating. In these warm, wet conditions, iron-rich minerals in the Martian soil

were transformed by chemical reaction with the abundant oxygen, undergoing an "oxidization" process that gave them their distinctive, rusty, red hue.

This early atmosphere was doomed to a slow deterioration over hundreds of millions of years, changing the planet's destiny. Weak Martian gravity was simply unable to hold onto the faster-moving, lighter gases such as molecular oxygen and nitrogen, both of which are considerably lighter than carbon dioxide. The absence of a magnetic field could also have played a role, permitting high-energy cosmic rays to bombard the upper atmosphere and splitting heavier molecules into lighter individual atoms that escaped more easily.

More dramatically, a major impact from space could have driven away much of Mars' ancient atmosphere in a single traumatic event. Whatever the cause, it is clear that most of Mars' atmosphere had disappeared by about 4 billion years ago, leaving the cold, exposed planet we know today.

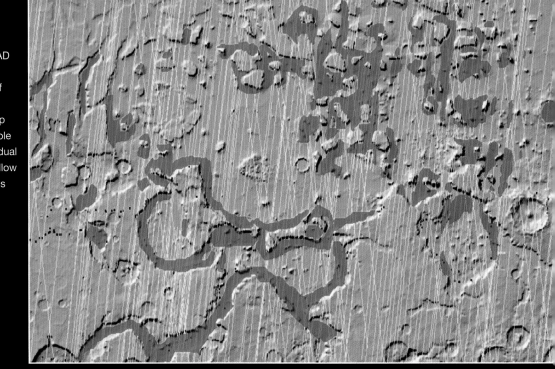

RIGHT This map of the mid-latitude Deuteronilus Mensae region, compiled using Mars Reconnaissance Orbiter's SHARAD subsurface radar instrument, highlights widespread deposits of ice linked to glacierlike features observed on the surface. The map was built up as MRO made multiple passes over the area (each individual "ground track" is marked by a yellow stripe), and reveals concentrations of ice up to half a mile thick. Ice seems to cluster at the base of steep cliffs and hillsides—most likely locations where it has been protected by overlying rock and dust from landslips.

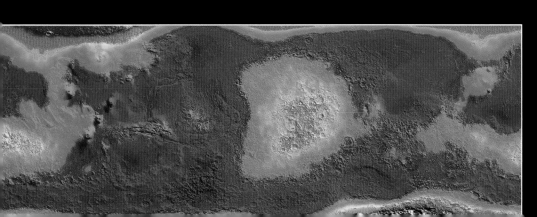

LEFT This map of Mars, compiled using the Neutron Spectrometer aboard 2001 Mars Odyssey, was the first evidence for widespread subsurface ice even at low latitudes. Color coding indicates the abundance of hydrogen in the soil, with even the least-enriched (blue) areas producing strong signatures for water.

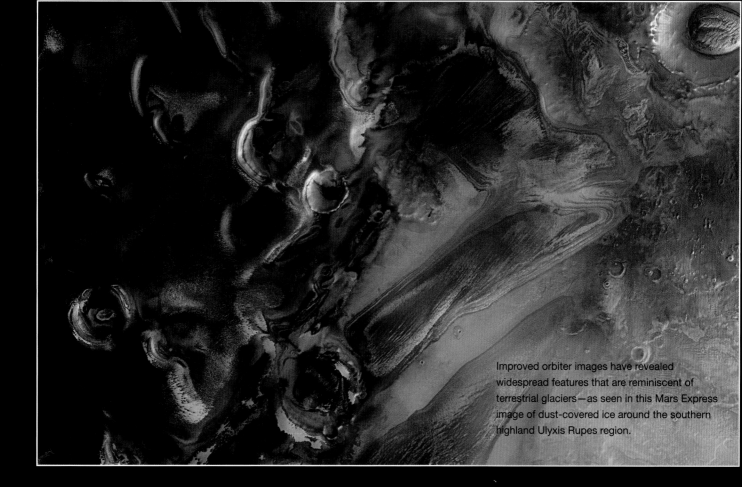

Improved orbiter images have revealed widespread features that are reminiscent of terrestrial glaciers—as seen in this Mars Express image of dust-covered ice around the southern highland Ulyxis Rupes region.

## Hidden Ice

A key question, of course, is, what happened to the Martian water as the planet slowly transformed into a cold desert? At one time, most planetary scientists assumed that it met a similar fate to the atmosphere. As the air pressure diminished, more and more water evaporated into gaseous form, leaving individual $H_2O$ molecules vulnerable to being split apart by cosmic rays. These ultimately blew away into interplanetary space as lightweight individual atoms. The residual water ice cap at the north pole was assumed to be all that remained.

That opinion began to change with the arrival of Mars Global Surveyor in late 1997 and the discovery that there was also water ice hidden beneath the permanent carbon dioxide of the south pole. Elsewhere, Surveyor's TES instrument detected the mineral olivine, interpreted as a sign that regions close to the equator had been water-free for many billions of years.

The Gamma-Ray Spectrometer aboard *2001 Mars Odyssey* made an even more dramatic breakthrough. Within months of beginning operations, mission scientists announced the detection of vast amounts of hydrogen enriching the upper layers of Martian soil. The most likely chemical compound to contain such large quantities of hydrogen is water ice.

The ice detected by *Odyssey* revolutionized ideas about the history of Mars. The greatest concentrations seemed to be around the south pole, but covered a broad swathe of

**ABOVE** Freshly formed craters, such as this one at a latitude of 64°N (formed at some point between March 2008 and May 2010) can blast away surface dust and reveal the pristine ice buried beneath. Over time, however, the newly exposed ice sublimates and the crater fades into the background.

the hemisphere up to mid-southern latitudes. The northern hemisphere also showed a large concentration of hydrogen in the Arcadia Planitia region (northwest of Tharsis), and lesser concentrations across the entire northern Vastitas Borealis and the low-latitude Arabia Terra. In places, the strength of the hydrogen signature implied that water ice was the dominant soil component. As William Boynton, principal investigator on the GRS instrument, put it, "It may be better to characterize this layer as dirty ice rather than as dirt containing ice."

**LEFT** One in a sequence of Mars Reconnaissance Orbiter images showing "recurring slope lineae" at mid-southern latitudes in Newton Crater. The precise mechanism driving the growth, darkening, and eventual fading of these features throughout the Martian seasons is still puzzling scientists.

To find the answer, scientists turned the high-resolution cameras of Mars Reconnaissance Orbiter toward the most obvious potential "smoking gun"—the gullies discovered by Mars Global Surveyor. Unfortunately, as we've seen, the discovery that these gullies undergo seasonal changes in the deep cold of winter and spring suggests that they are actually created by the evaporation of carbon dioxide frosts rather than the flow of water.

MRO's cameras revealed another type of feature occurring in similar areas to the gullies—dark streaks that are just a few feet wide or less, far smaller than the gullies, but may extend for hundreds of yards. In contrast to the gullies' preferred location on colder, pole-facing slopes, these "recurring slope lineae" (RSLs) are found at mid-southern latitudes on the warmer slopes that face the equator. While, like the gullies, they also undergo seasonal changes, the RSLs grow longer and darker during Martian summer before fading over the winter. All the evidence suggests that RSLs are shaped by a process that requires considerably higher temperatures than gully formation, and occur in conditions that are too warm for solid (or liquid) carbon dioxide to be playing a role.

The only viable alternative is water, perhaps a salty brine with a lower freezing point than pure $H_2O$. The discovery of salt deposits across mid-southern latitudes and elsewhere suggests that such brines were certainly widespread in the past, but water has not yet been directly detected on the RSLs, and their dark appearance cannot be explained simply by them being wet. Instead, the most popular theory so far suggests that the flow of brine either on or just below the surface during summer creates rougher streaks that appear dark, but the big question with this model is, what process then smoothes and lightens the terrain in winter?

## Highland Glaciers

Far from evaporating and being blown away into space, much of Mars' ancient water instead retreated below the surface, where it still remains in deep-frozen storage. Once this initial breakthrough was made, the discoveries kept coming. In 2005, ESA's Mars Express captured images of a disk of white ice on the floor of a crater some distance from the north polar ice cap, the first of several such icy outcrops to be discovered.

Even more impressive was the discovery of a "frozen ocean" at equatorial latitudes on the Elysium Planitia. Here, fragmented blocks of material that have rotated relative to each other and are set within a slightly lower plain resemble patterns of pack ice seen on Earth. Since the arrival of Mars Reconnaissance Orbiter, even more icy features have come to light, including glaciers in the southern highlands and around the huge impact basin of Hellas Planitia.

Based on all these satellite observations, and confirmed by the 2008 measurements of NASA's Phoenix polar lander, it now seems clear that much of the visible Martian surface is dusty ice rather than solid rock. The next question is whether the ice ever melts, either below ground or at the surface.

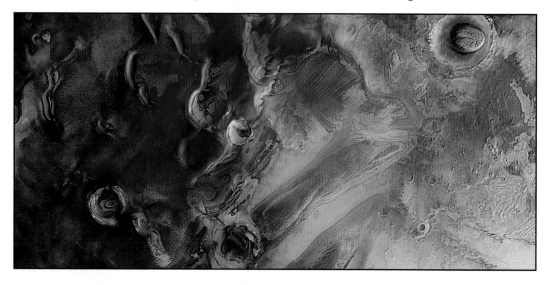

**LEFT** The HRSC stereoscopic camera on Mars Express is not just used for generating spectacular perspective images. It can also be used to generate digital terrain models—color-coded maps of surface elevation. This example shows the same area of Ulyxis Rupes shown in true color on the previous page.

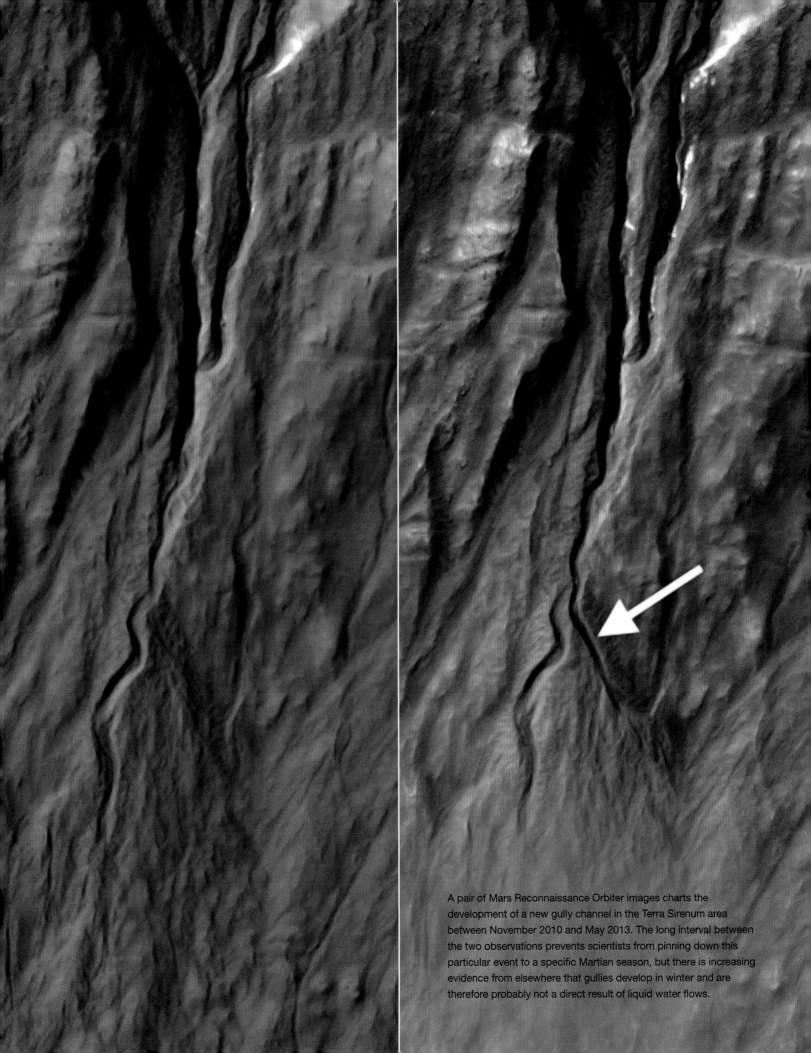

A pair of Mars Reconnaissance Orbiter images charts the development of a new gully channel in the Terra Sirenum area between November 2010 and May 2013. The long interval between the two observations prevents scientists from pinning down this particular event to a specific Martian season, but there is increasing evidence from elsewhere that gullies develop in winter and are therefore probably not a direct result of liquid water flows.

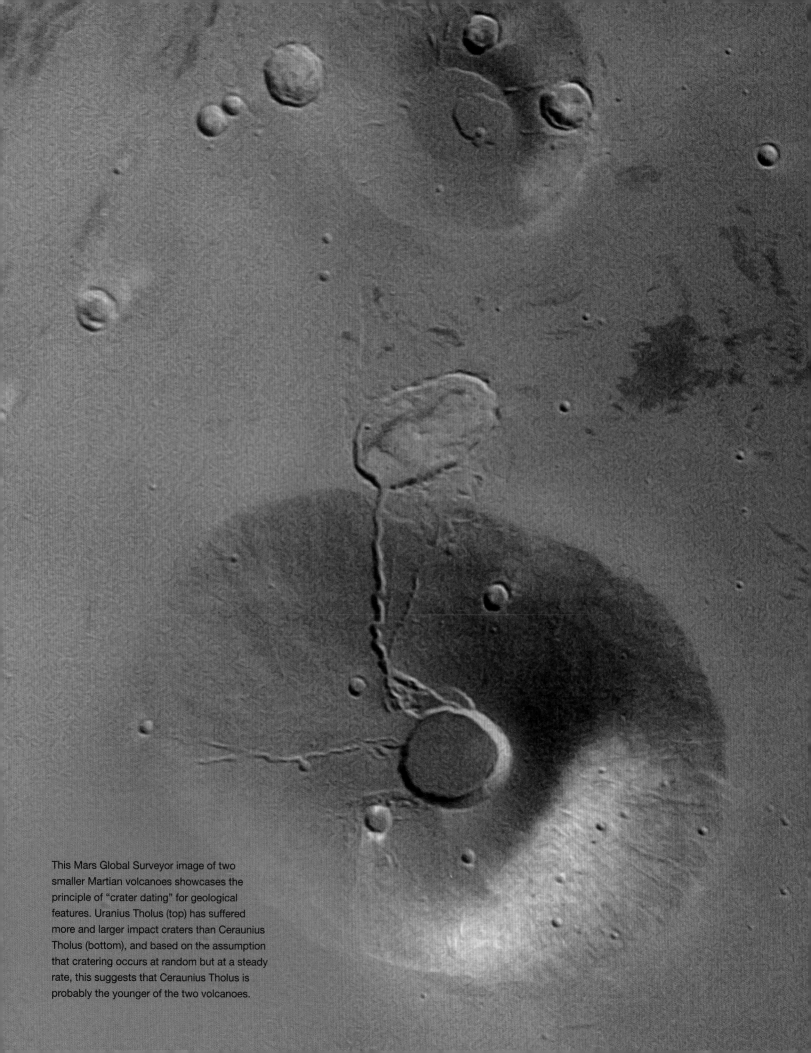

This Mars Global Surveyor image of two smaller Martian volcanoes showcases the principle of "crater dating" for geological features. Uranius Tholus (top) has suffered more and larger impact craters than Ceraunius Tholus (bottom), and based on the assumption that cratering occurs at random but at a steady rate, this suggests that Ceraunius Tholus is probably the younger of the two volcanoes.

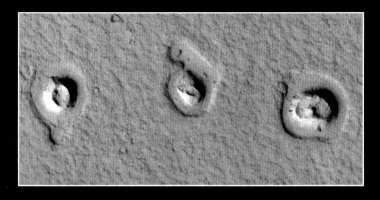

ABOVE So-called "rootless cones" are small volcanoes formed when lava interacts with subsurface ice or water. The discovery of many such volcanoes with little crater damage is some of the best evidence for recent volcanic activity on Mars.

## An Active Planet?

Just as exciting as the potential discovery of water on the Martian surface are the controversial reports of methane in the atmosphere. This volatile chemical (formula $CH_4$) was first detected in spectroscopic measurements of Mars around 2004, and apparently confirmed by instruments aboard the Mars Express probe. Methane is intriguing because it is a very short-lived chemical compound in Martian conditions. Bombarded by radiation from the Sun, it should break down in a matter of centuries to form carbon dioxide.

In other words, if methane exists on Mars today, then it cannot simply be a remnant from ancient times—it must be continuously regenerated. That's interesting because the most plausible mechanisms for producing methane involve active volcanism, hydrothermal activity, or the biological processes of certain microbes.

Interest in the possible methane signature stepped up a notch in 2009, when researchers at NASA's Goddard Space Flight Center, who had continued the original telescopic studies, announced that they had identified distinct localized plumes of methane in the atmosphere, and even seasonal rise and fall. This raised as many questions as it answered. Any new methane generated should naturally degrade over a matter of centuries, but it's hard to come up with a mechanism to explain its decline in a matter of months.

In 2012, another team from NASA's Ames Research Center put forward plausible arguments that the methane detection was an illusion, a false result caused by detection of gas in Earth's own atmosphere, coupled with misinterpretation of the Mars Express results.

## Rising Methane Readings

Initial results from the *Curiosity* rover in 2012 and 2013 seemed to settle the matter, as the rover's instruments effectively ruled out any significant atmospheric methane at the landing site. But then, in December 2014, NASA reported Curiosity's detection of a sudden methane "spike,"

building up over a matter of months until the gas comprised 7 parts per billion in the atmosphere. It seems that methane on Mars is indeed a reality, but the mechanisms that create (and destroy) it are still open to question.

Continuing volcanic activity is one possibility, and there is evidence for young volcanoes in some region of Mars. Short-lived "outgassing" events might seem to fit the pattern of sudden methane spikes, but there's not yet any conclusive link between localized concentrations of methane and likely volcanic regions.

Hydrothermal action, in particular a chemical reaction known as *serpentinization* that generates methane when warm water comes into contact with olivine minerals, is another possibility, and this might also be linked to underground volcanic activity keeping the water warm.

The third option, methane-producing microbes, is the most exciting of all. Localized outbursts could be explained by their response to seasonal changes in temperature and frost cover, and a "biogenic" source should, in theory, leave a telltale signature in the isotopic composition of the methane. In other words, it should contain more lightweight carbon-12 atoms than might otherwise be expected. For the moment, pending new measurements and future missions to Mars, there is no definitive answer to this tantalizing question.

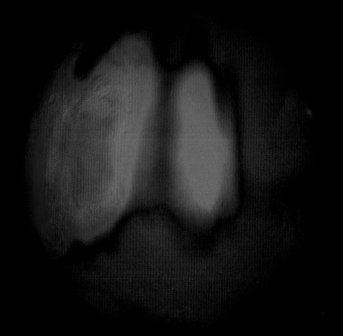

ABOVE This map, based on observations from NASA's Infrared Telescope Facility and the Keck Telescope, both on Hawaii, charts the concentration of methane in the Martian atmosphere during northern summer. For Mars scientists and enthusiasts, the discovery and confirmation of methane is one of the most exciting developments of recent times.

# CURIOSITY AND ITS KIN

Given the difficulties of sending astronauts to Mars, the next best alternative is to send a mobile robot capable of moving over the surface and mimicking at least some human abilities. Since 1997, four such robot vehicles have explored the Red Planet, delivering a wealth of new knowledge and a taste of conditions on the surface of the Red Planet.

## Tethered Rovers

The idea of a robot rover on Mars dates back to the very beginnings of the Space Age. Soviet engineers equipped the very first Mars landers, Mars 2 and 3, with small vehicles called Prop-M rovers.

BELOW An artist's impression of NASA's car-sized Curiosity rover, the most sophisticated space probe ever launched, on the surface of Mars.

LEFT The Curiosity rover sits in Gale Crater surrounded by a maze of its own tracks, with the rising slope of its ultimate target, the outcrop known as Mount Sharp, behind it. Rovers such as Curiosity have added immeasurably to our understanding of Mars and offer the closest thing to sending astronauts that we can achieve with current technology.

Had the craft successfully landed and maintained contact with Earth, their job would have been to explore the surface around their landing sites, moving on skilike tracks at the end of a 50-foot (15-m) umbilical tether.

Following the failure of both these and other early missions, such ambitious ideas were set aside to concentrate on the challenges of simply reaching the Red Planet. As a result, the first rovers to land on another world were the Soviet Lunokhod Moon missions of the early 1970s. Aside from a fairly basic robot arm for scooping up surface rocks and dust, NASA's Viking landers of 1976 were obstinately static.

It was only when NASA turned its attention back to Mars in the 1990s that rovers once again became a priority. Following the failure of the Mars Observer mission, the agency's plans for Mars were unified under a single Mars Exploration Program in 1993. Tasked with developing new technologies and paving the way for manned missions, the program planners also had to come to terms with NASA's new general goal of "faster, better, cheaper" interplanetary missions.

If rovers were to be the way ahead, then it made sense to start with a "proof of concept" mission that could be built quickly on a limited budget, and so the Mars Pathfinder mission was born.

Ares Vallis to the south of Chryse Planitia, which promised a relatively smooth terrain with an interesting variety of material washed down from the highlands.

Pathfinder's descent tested several technologies that would prove useful on future missions. As it plunged into the atmosphere, a conical heat shield protected the lander from friction with the gas, and helped it slow from an entry speed of almost 4 miles per second (6 km per second) to a merely supersonic 833 mph (1,340 km/h). At this stage, a parachute deployed from the upper protective "backshell," further slowing the descent. The heat shield was now ejected, and the probe itself lowered from the backshell on a 65-foot (20-m) tether.

When the probe was around 1,150 feet (350 m) above the ground, according to its onboard radar, protective airbags inflated around it. Finally, at an altitude of just over 320 feet (98 m), three retrorockets mounted in the backshell fired, slowing the descent speed to zero about 65 feet (20 m) up. Pathfinder was then released from its tether, bouncing to a halt on its cushion of airbags. The final stage in the landing sequence was to deflate the bags in sequence to set the spacecraft upright on the ground.

## Sojourner Truth

The Pathfinder mission involved two separate elements— the pyramid-shaped lander and base station, and the rover contained within. Once safely on the surface, the lander opened three petal-like side sections to expose solar panels, and raised a pole carrying its two main experiments, a stereoscopic camera for taking three-dimensional images, and a weather station. Opening the lander petals also allowed the Sojourner rover to roll out onto the Martian surface. Named after 19th-century anti-slavery and womens' rights campaigner Sojourner Truth, this six-wheeled vehicle was 25.5 inches (65 cm) long and 19 inches (48 cm) wide, with a flat upper surface covered in solar panels. Weighing 25 lb

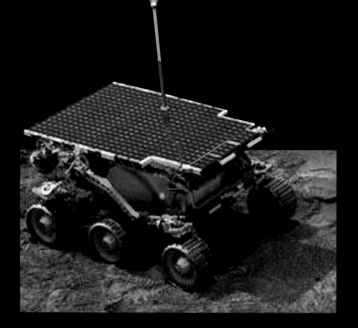

**ABOVE** Mankind's first rover on another planet, the plucky Sojourner, was little larger than a remote-controlled toy tank—yet it massively outlived its planned mission.

## Pathfinder on Mars

Pathfinder was launched from Cape Canaveral on December 4, 1996, arriving at Mars on July 4, 1997. Developed at NASA's Jet Propulsion Laboratory in just three years, the spacecraft cost a mere $150 million. The total budget, including launch on a Delta II rocket and Earth-based mission operations came to $280 million—less than one-tenth of a Viking mission after adjustment for inflation.

With no operational spacecraft in orbit, mission scientists were forced to choose a landing area based on Viking images from

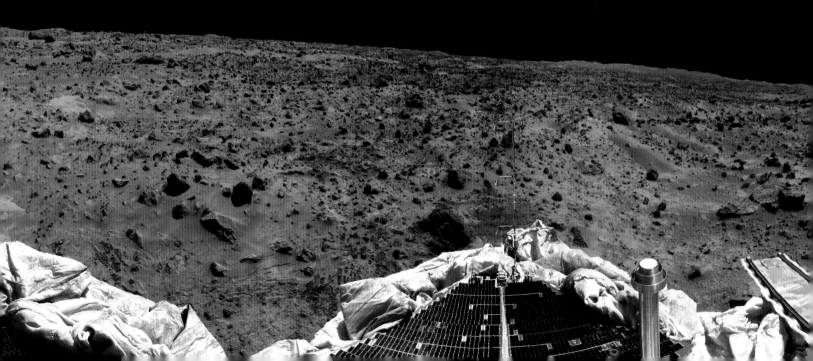

(11.5 kg), it carried two monochrome cameras mounted at the front, and a color one at the rear.

Sojourner's other main experiment was the Alpha Proton X-ray Spectrometer (APXS), an ingenious device built by a German–U.S. team. The spectrometer carried a small source of alpha radiation on board, which could be directed as a beam onto the surface of any rocks the rover encountered. This would cause the elements within them to emit X-rays of different energies, producing a "fingerprint" by which individual atoms and molecules could be identified.

## Virtual Reality

One of the rover's key aims was to test new navigation methods, helping engineers get to grips with the challenges of driving on a planet where, thanks to the limited speed of radio communications, the round-trip time from sending a command to the vehicle's wheels to seeing the response back at Mission Control was at least three minutes, a huge lag compared to the delay of a couple of seconds for the Lunokhod Moon rovers.

To overcome this, engineers created software that was capable of limited autonomy, finding the best path to an assigned target or way-station, and avoiding unexpected hazards along the way. The front-mounted cameras were a vital element of this system. Their images were combined back on Earth in a virtual reality headset. Despite this, Sojourner's top speed was limited to less than half an inch (1 cm) per second, and it never strayed more than 40 feet (12 m) from its parent spacecraft, upon which it relied as a radio relay to Earth.

Ambitions for the Pathfinder mission were limited. Engineers hoped that the rover might last for seven days and the base station for 30. In the end, both elements far surpassed these predictions, with communication finally failing on September 27, after 83 Martian days, and Sojourner remaining active right up to the end. In total, the rover traversed more than 330 feet (100 m) of Martian terrain, sending back more than 500 images and analyzing 16 separate rocks.

**ABOVE** Sojourner applies its APXS spectrometer in an area nicknamed the "Rock Garden," where rocks appear to be tilted in a "downstream" direction by ancient water flows.

**BELOW** This 360-degree panorama is a mosaic of images from the camera aboard the main Pathfinder spacecraft. The lander's solar-panel petals can be seen spread out around it, with the distant hills known as "Twin Peaks" in the center background, and Sojourner investigating a rock at right.

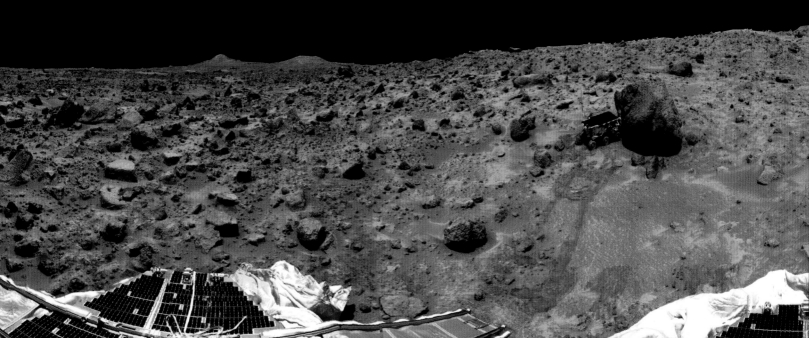

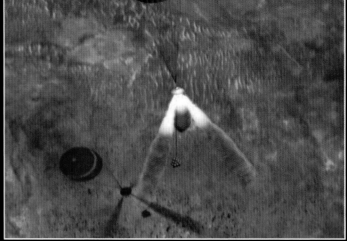

ABOVE Following an initial descent slowed by hypersonic parachute, retrorockets in the backshell fire about 70 feet (20 m) above the surface, slowing the spacecraft's descent speed to zero.

ABOVE The Delta II rocket carrying MER-B, Opportunity, takes off from Pad-17B at Cape Canaveral in a spectacular night launch on July 7, 2003.

## Mars Exploration Rovers

By any standards, the Pathfinder mission had been an outstanding success, so the way was clear for a more ambitious second generation of rovers. For the 2003 launch window (the closest opposition in recent history), NASA planned to send a pair of missions, dubbed the Mars Exploration Rovers. Officially designated MER-A and MER-B, the two rovers were named Spirit and Opportunity shortly before their launch—as for Sojourner, names suggested by the winner of a student essay competition.

The rovers were much larger than Sojourner. They were 5 feet 3 inches (1.6 m) long and 7 feet 6 inches (2.3 m) wide thanks to their fan-like arrangement of solar panels. A mast carrying cameras and other equipment gave each a total height of

5 feet (1.5 m), while a robot arm extended their reach and sampling ability. A high-gain (radio dish) antenna permitted direct contact with Earth at high data rates, while an omnidirectional (low-gain) antenna allowed data to be sent at a slower rate via Mars-orbiting space probes, and could also be used for direct communication with Earth in an emergency. This did away with the need for a base station to act as a radio relay, but the rovers were still sent to Mars in pyramid-shaped landers with more than a passing resemblance to Pathfinder. In fact, the entire landing procedure was essentially a scaled-up version of that used on the earlier mission, modified to cope with a payload weighing more than 397 lb (180 kg).

## Twin Missions

In an echo of the Mariner and Viking philosophy, each rover carried an identical suite of instruments, an insurance policy in case of failure that would double the project's scientific yield if all went well. Instruments were either carried on the rover's elevated mast, or mounted on the versatile robot arm, known as the "instrument deployment device" (IDD).

Mast experiments included a high-resolution, full-color panoramic camera known as Pancam, and a monochrome Navcam, a suite of two paired cameras mounted at different angles, producing three-dimensional images to aid navigation. A third instrument was the Miniature Thermal Emission Spectrometer (Mini-TES), an infrared spectrometer operating on the same principles as Mars Global Surveyor's TES experiment, and capable of analyzing the composition of rocks from a distance.

Meanwhile, the IDD arm carried a camera of its own, the Microscopic Imager, alongside an Alpha Particle X-ray Spectrometer similar to that carried on the _Sojourner_ rover, and another advanced tool for analyzing minerals called a Mossbauer Spectrometer. In addition, the IDD carried a Rock Abrasion Tool for sweeping aside dust and grinding down

ABOVE The delivery vehicle, cushioned in airbags, is released from the tether attaching it to the backshell, and bounces to a halt.

ABOVE Once it has come to a halt, the airbags deflate in sequence to put the delivery vehicle upright. Its three sides now open to release the rover onto the surface.

exposed surfaces, and magnets for collecting and analyzing magnetic particles in the soil (such as those formed from iron-rich minerals).

As with Sojourner, each Mars Exploration Rover had a six-wheeled design, with each 10-inch (250-mm) diameter wheel powered by its own motor and therefore capable of independent operation. An ingenious suspension method known as a "rocker-bogie" system ensured that all wheels remained in contact with the ground even on rough terrain, and advanced hazard-avoidance systems were built into the rover's software, using data from paired Hazcam cameras at the front and back of the vehicle chassis, together with that from the elevated NavCam. Nevertheless, the rovers still moved at a cautious top speed of 2 inches (5 cm) per second. Engineers planned for the missions to last at least 90 days on the Martian surface, during which time the vehicles would use a great deal of power. The operating power was mainly drawn from the solar panels, but strategically positioned radioisotope heater units, generating heat from the decay of a small amount of radioactive material, kept vital systems above a critical temperature of -40°F (-40°C).

An artist's impression sh
one of the Mars Explora
Rovers at work on the su
of Mars, using the array
instruments on its robot
to inspect a nearby rock

Launched on June 10, 2003, MER-A (usually known simply as the Spirit rover) touched down on Mars almost seven months later on January 4, 2004. Its landing site was at the center of a large crater called Gusev, 14 degrees south of the Martian equator. Orbital surveys suggested this 103-mile (166-km) crater, on the boundary between southern highlands and northern plains, had once been filled by water running northward along a channel known as Ma'adim Vallis.

There was even what looked like a dried-up lake bed in the middle, and NASA scientists and engineers deliberately chose this target as a good place to look for sedimentary rocks laid down in the Red Planet's warmer, wetter past.

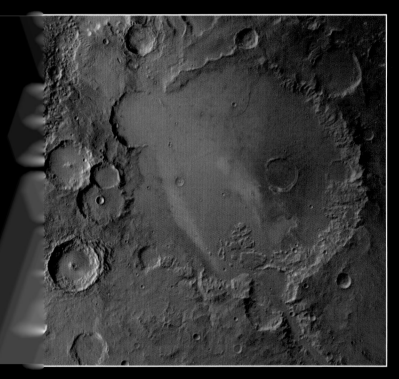

ABOVE This mosaic of images from *Mars Odyssey's* THEMIS infrared instrument shows Spirit's landing site in Gusev Crater. The inlet from Ma'adim Vallis, which once carried water here from the southern highlands, lies near the bottom of the picture.

to a range of low hills. The landing site was named Columbia Memorial Station in honor of the Space Shuttle that had broken up on re-entry to Earth's atmosphere barely a year before. Seven of the nearby Columbia Hills were individually named after the seven astronauts killed in the tragedy.

Spirit's first target for investigation, and testing of its instruments, was a nearby shallow depression nicknamed Sleepy Hollow. But just 17 sols (Martian days) into the mission, Spirit's transmissions suddenly fell silent. In emergency transmissions the following day, and over the next 15 days, the engineers painstakingly traced the problem to a memory error and nursed the rover back to full health.

NASA now set Spirit on a long trek toward the Columbia Hills in search of interesting geology, pausing to look at various rocks along the way. Initial analysis suggested these rocks were volcanic in origin, but had perhaps been modified by the action of water after their formation, although it now seems likely that the crater lake deposits that may once have covered Gusev have since been buried beneath later volcanic material.

The rover passed by the 656-foot (200-m) Bonneville Crater and 328-foot (100-m) Missoula Crater without detailed investigation, but had already long outlasted its planned 90-day mission by the time it reached the foot of Husband Hill on Sol 159.

## The Pot of Gold

Nearby, within a small and slippery depression, scientists singled out an interesting-looking rock nicknamed Pot of Gold for further investigation. Although reaching the rock was difficult, it proved worthwhile—analysis with the APXS instrument in June 2004 identified hematite, an iron-rich mineral, which on Earth forms in the presence of water. Halfway around the planet, similar discoveries were also being made by Opportunity.

As the mission continued into late 2004, the next major challenge facing Spirit was the onset of autumn, with colder nighttime temperatures and a lower, weaker Sun generating less energy through the increasingly dusty solar panels. Seeing Spirit through the winter involved implementing a range of special measures, including the introduction of a

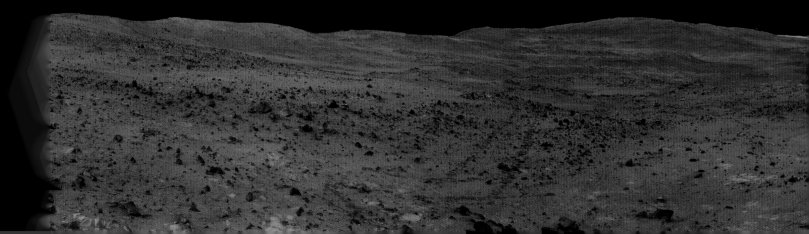

**RIGHT** 16 sols into its mission, Spirit snapped this image of its now-redundant delivery vehicle on the Martian surface. The Columbia Hills, soon selected as Spirit's main target for investigation, lie in the background.

power-saving "deep sleep" at night, despite the risk of losing individual instruments to the cold.

Spirit's routes now had to be planned specifically to keep the rover's upper surface angled toward the Sun during its ascent of Husband Hill. Despite careful steering, the chosen routes proved unexpectedly hazardous, and in February 2005 it was decided to send a significant upgrade to the rover's software that would allow the vehicle to make more autonomous driving decisions.

Despite the risks, Spirit survived its first Martian winter in good shape, and in March 2005 received an unexpected boost when the slow decline of its solar panels reversed literally overnight. Previously the cells had been causing some concern as they were now operating at just 60 percent of

their original efficiency. It came as a welcome surprise when they suddenly jumped to 93 percent.

The cause of this power boost soon became clear when Spirit spotted dust devils passing close by. It seemed that one of these Martian twisters had passed directly over the rover, dislodging the layers of accumulated dust but leaving the vehicle itself unharmed. This was the first of many such "cleaning events" to benefit both rovers, and vastly extended their operational lifetimes.

**BELOW** Descending from the summit of Husband Hill in November 2005, Spirit paused to capture this detailed panorama of the Columbia Hills. The complete image is stitched together from 405 individual photographs.

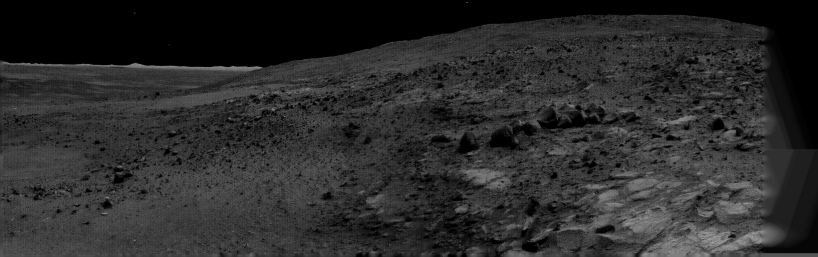

ABOVE A furrow created by Spirit's faulty wheel dragging behind the rover unexpectedly led to one of the rover's greatest discoveries—a large patch of bright, sandy silica that may have formed in an ancient hot spring.

## Microbial Life?

As Spirit continued its investigation of Husband Hill, it spotted a distant outcrop of interesting-looking rock that was soon named "Home Plate." Investigation had to be combined with a plan to reach McCool Hill, another Sun-facing slope where it could wait out both the next Martian winter in late 2006, and a solar conjunction that limited communications with Earth.

Home Plate turned out to be a particularly rich target, since it proved to contain carbonate rocks. These minerals form readily from the interaction of rock, water, and atmospheric carbon dioxide, but also dissolve rapidly in acidic waters. Although there are a few rare Earth microbes that thrive in such hostile conditions, this was the first confirmation of an ancient Martian environment with the potential to support more common Earth organisms.

While driving to its winter refuge, Spirit hit its first incurable problem when a front wheel, which had been jamming intermittently for some time, finally stopped turning altogether. Here the rover's versatile design proved useful, as it was able to carry on driving in reverse, dragging its lame wheel behind.

In fact, as Spirit got under way again in spring 2007, this fault led to an important and unexpected discovery when scientists spotted an intriguing patch of light-colored dust exposed by the dragging wheel.

Close analysis with the APXS instrument revealed that it was rich in silica (the glassy component of most Earth sands), and was probably formed in an ancient hot-spring environment that would have been ideal for microbial life.

## Hibernation Mode

That summer, as Spirit returned to the Home Plate area, it found itself overwhelmed by a major global dust storm. With dust both darkening the skies and falling onto its solar panels, Spirit's power ran dangerously low, forcing it into hibernation mode to preserve the remaining power in its batteries.

The approach of another winter left engineers with little choice but to risk maneuvering the rover onto a Sun-facing slope, but conditions worsened with a winter dust storm in late 2008, and auxiliary heating to the instruments such as Mini-TES was reluctantly shut down to save all possible power.

Matters improved a little after the solar conjunction of early December 2008. Communications were re-established and a series of cleaning events restored the solar cells to about 60 percent of their original efficiency. Spirit was on the move again, but just as things seemed bright for another summer of exploration, disaster struck.

## End of the Mission

On May 1, 2009 (Sol 1892 of its mission), Spirit became bogged down in a "sand trap" of slippery iron sulfate minerals that had been hidden beneath normal-looking dust.

On Earth, computer models were used and a lighter weight mock-up of Spirit was built in order to conduct tests

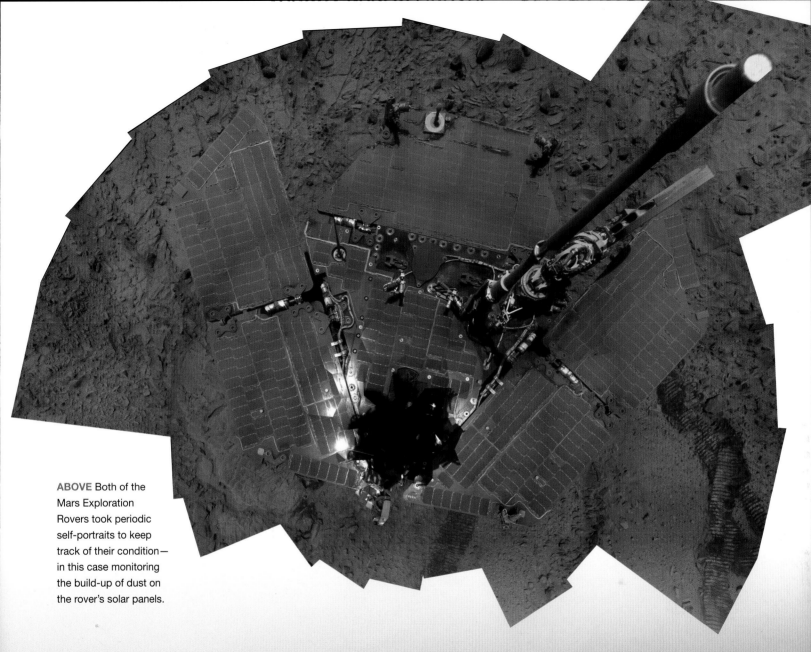

ABOVE Both of the
Mars Exploration
Rovers took periodic
self-portraits to keep
track of their condition—
in this case monitoring
the build-up of dust on
the rover's solar panels.

simulating the low-gravity environment, weak atmospheric
pressure, and the soil conditions the rover was experiencing.

After several months of intensive planning, recovery
maneuvers later in the year cost the use of another wheel, and
it was clear that Spirit was now thoroughly embedded at a

location named Troy. Unable to re-orient the rover for the
coming winter, scientists made the best of the now-stationary
Spirit's last few months of life, until communication was finally
lost due to dwindling power on March 22, 2010 (Sol 2210).
Attempts to re-establish contact the following Martian summer
proved fruitless and the mission was terminated in May 2011.

BELOW This stunning view captures the moment of sunset over
Gusev Crater on Sol 489 of its mission (May 19, 2005). Sunset and
sunrise images are not just beautiful—they also help scientists to
better understand the structure of the Martian atmosphere.

# Opportunity Rover

Officially designated MER-B, the Opportunity rover was launched four weeks after its sibling, on July 7, 2003, but thanks to the changing relative positions of Earth and Mars, reached the Red Planet just three weeks after Spirit.

It parachuted into the Martian atmosphere on January 25, 2004, without problems, but came to rest some 16 miles (25 km) east of its intended landing site on the flat equatorial plain known as Meridiani Planum. What was more, Opportunity's protective airbags rolled it into a small impact crater, so that, when the balloons were deflated and the rover deployed, the first images showed a shallow ridge surrounding it on all sides.

This happy accident offered a wealth of targets for the first phase of Opportunity's mission. The site, subsequently named Eagle Crater, was a shallow bowl some 72 feet (22 m) in diameter, with an outcrop of light-colored, layered rocks emerging from the darker soil along its northwestern wall.

Such rocks must have formed from the accumulation of fine sediments, either underwater or in layers of volcanic ash, and close-up photos soon revealed features that suggested the action of flowing water. One in particular, named El Capitan, was singled out for detailed study.

The Rock Abrasion Tool was used to grind away a section and look at the internal structure, while the Mini-TES instrument and the Mossbauer Spectrometer were used to analyze its chemical constituents, revealing sulfate chemicals and a mineral known as jarosite, both signs that the rock had formed underwater. Spherical pellets rich in the iron compound hematite were also found in the crater—a further indication that the underlying rocks had formed underwater.

## On to Endurance

With Opportunity's unexpected surroundings now mapped from above by Mars Global Surveyor, scientists selected a larger nearby crater, Endurance, as the rover's next target.

Eleven times the diameter of Eagle, Endurance had deeper and more complex layers of rock exposed in its steeper walls. Mission controllers decided to risk driving the rover into the crater, despite the risk that it might not get out again. Test drives down the crater slope in June 2004, however, showed that the walls were surprisingly stable, capable of supporting

Opportunity on gradients of up to 30 degrees. Opportunity spent the next six months carrying out a detailed investigation of the crater before climbing back out and continuing its journey across the plain.

In January 2005, while close to its own discarded heat shield, it identified a strange rock that proved to be the first meteorite ever found on the surface of another planet.

## Bogged Down

The mission's next major target was to be the even larger Victoria Crater, some 2,400 feet (730 m) in diameter, and a daunting 4 miles (7 km) south of the landing site. But in April 2005, Opportunity unexpectedly stranded itself in a small sand dune, with all four corner wheels deeply embedded in the Martian dust.

Extricating the rover was a painful process, involving detailed testing on Earth, followed by delicate maneuvers on Mars. For six weeks, the mission controllers pored over computer models and physical test results before coaxing Opportunity an inch at a time out of the dune. The rover's precarious situation was thoroughly re-analyzed after each intricate move, and Opportunity was finally freed from the aptly named Purgatory Dune on June 4.

## Heading South

Opportunity set off south once again, the team now wary of the terrain on Meridiani Planum, which was proving to be a rather different environment from other Martian landing sites. There were few scattered rocks to present hazards, but plenty of rippling dunes.

In an effort to avoid any further soft ground incidents, engineers gave Opportunity a software upgrade designed to keep a closer check on the traction of its wheels, and stop it digging itself into another sand trap. The rover then headed for a large, shallow crater called Erebus where it would explore between October 2005 and March 2006 before carrying on toward Victoria Crater.

BELOW This panoramic view shows an area known as Erebus Rim, along the edge of a heavily eroded, ancient crater. The distinctive pattern of cracked rock emerging from the soil is known as "etched terrain."

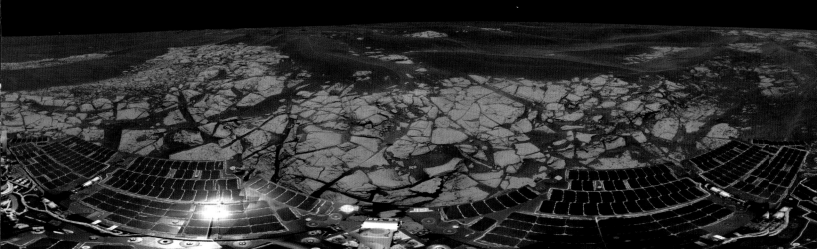

### Down in Duck Bay

After a long drive that included a hazardous dust storm, and several "cleaning events" (likely encounters with Martian dust devils that lifted the accumulated dust from its solar panels), Opportunity arrived at the edge of Victoria Crater in September 2006—Sol 951 of its mission.

Over the next few months, the rover inspected the northwest edge of the crater rim, identifying a way down to the crater floor using a shallow slope in a scallop-like alcove named Duck Bay. Between June and August 2007, however, Opportunity was threatened by the same global dust storm that menaced Spirit on the other side of the planet.

Power in Opportunity's batteries dropped to critical levels, and it was only in late August that the storm abated, skies began to clear, and the solar panels were able to recharge. The rover finally made its way down into Duck Bay in September 2007, spending most of the next year inspecting the interior of the crater wall, although it hit another problem in April 2008 when its robot arm malfunctioned.

The arm's shoulder joint had been problematic, periodically overheating and locking solid, since the second day on Mars. Engineers had successfully managed the issue for some 1,500 sols but the arm now completely refused to move, remaining stowed in its protective "tucked-in" position.

The solution involved deliberately overheating the shoulder in order to unstick the joint, moving it one last time into a position where the arm instruments could be used. Engineers had to modify their techniques for driving the rover in order to cope with the hazard of a permanently extended arm.

### The Drive to Endeavour

Leaving Victoria Crater in late August 2008, Opportunity's next target on its tour of Martian craters was the even larger Endeavour, almost 14 miles (22 km) in diameter.

The drive across 7.5 miles (12 km) of plain was expected to take about two years, but in May 2010, engineers revised the route to 12 miles (19 km) in order to avoid a potentially dangerous dune field. Along the way, the rover inspected

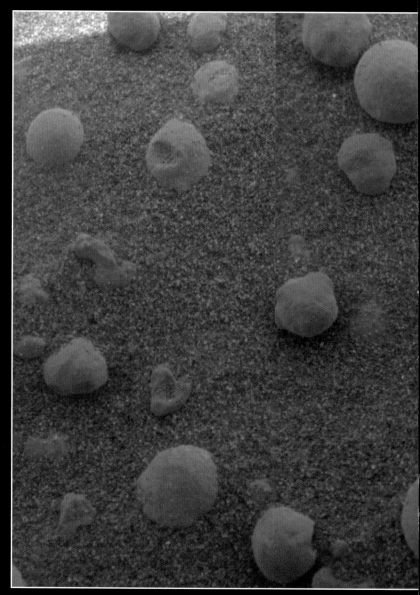

**RIGHT** A close-up view of the hematite-rich spherules Opportunity discovered scattered near its landing site in Meridiani Planum. Many are clearly embedded in the surrounding soil while others are sitting on the surface, suggesting that they formed within rock that has since eroded away.

**BELOW** A Mars Reconnaissance Orbiter image of the half-mile (730-m) wide Victoria Crater. Opportunity explored the crater rim and interior at great length, focusing on the area around Duck Bay (about the ten o'clock poasition in this image).

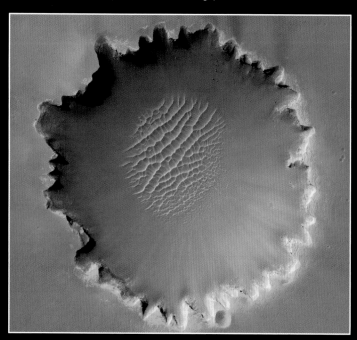

several meteorites and a large chunk of ejecta—rock flung out from deep in the Martian crust by a meteorite impact.

Opportunity finally arrived at the rim of Endeavour in early August 2011, on Sol 2709. Unlike at Victoria, a route was planned to explore interesting features along the edge without entering the crater itself. Endeavour's eroded crater seems to expose some rocks that are up to 4 billion years old.

One early discovery was the Homestake formation, a vein of light-colored rock that proved to be gypsum (hydrated calcium sulfate), a mineral that forms only in the presence of water. Homestake was probably an ancient Martian spring where underground water flowed through rock fractures.

Opportunity celebrated its tenth birthday on Mars in early 2014, and continues to examine features on the Endeavour Rim, including its oldest rocks yet. The rover has now traveled across more than 25 miles (40 km) of Martian landscape and, at the time of writing, is still in remarkably good shape after 11 years on the Red Planet.

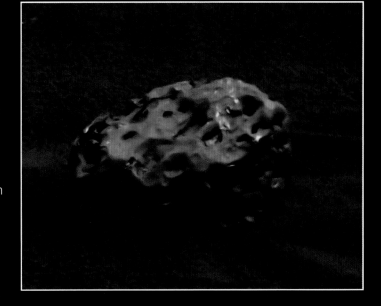

**ABOVE** The first meteorite discovered on another planet, Heat Shield Rock is a roughly football-sized lump of iron and nickel that probably formed inside an asteroid that fragmented long before it landed on Mars.

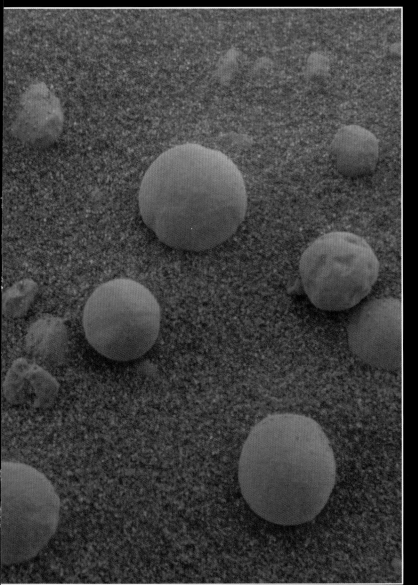

The Martian spherules' nickname of "blueberries" comes from their apparent hue in false-colored images used by NASA to highlight their presence.

This image, taken by Phoenix shortly after landing, shows a pebble-strewn landscape with distinctive polygonal cracks caused by expansion and contraction of the Martian permafrost.

## Phoenix on Mars

While Spirit and Opportunity both far outlasted their planned missions, NASA's next lander on Mars was doomed to a short life from the outset. Phoenix was a successor to the ill-fated Mars Polar Lander, lost in 1999, but unlike that mission, which was to target the plains around the south pole, Phoenix aimed to land on the northern Vastitas Borealis.

The probe was an immobile lander with an appearance that harked back to the Viking missions. Based on a spacecraft that had been mothballed following the cancellation of the Mars Surveyor 2001 mission, it carried variants of several instruments from the earlier Polar Lander.

Launched from Earth on August 4, 2007, it touched down at a latitude of 68 degrees North on May 28 the following year, late in the northern hemisphere spring. Unlike the complex balloon descents made by previous rovers, Phoenix used a simpler solution, with retrorockets mounted beneath the spacecraft.

## Searching for Ice

In order to generate sufficient electricity using solar power at such a high latitude, Phoenix deployed two large circular arrays shortly after landing.

With a total area of over 31 square feet (2.9 m²), more than twice the collecting area of the Mars Exploration Rovers' panels, the lander automatically oriented itself on an east-west line for maximum exposure to the Sun.

One of the mission's primary goals was to confirm the presence of water ice beneath the Martian soil at high latitudes, as suggested by orbital observations from *2001 Mars Odyssey* and other probes.

Although the spacecraft itself was static, it used a robot arm to scoop up soil samples and pass them to various experiments. Instruments on board included the Thermal

and Evolved Gas Analyzer (an oven and mass spectrometer for analyzing the chemical content of soil samples), the Thermal and Electrical Conductivity Probe (a series of four probes that measured properties of the soil), and the Microscopy, Electrochemistry, and Conductivity Analyzer (a laboratory to study aspects of soil chemistry).

The lander also carried a meteorology station and three cameras—the Descent Imager for photographing the landscape during final approach, the pillar-mounted Surface Stereo Imager, and a third Robotic Arm Camera.

## Retro Rocket Effect

Phoenix set down on a flat plain strewn with pebbles, whose major feature was a network of polygonal troughs in the soil, measuring several feet across. Comparing these formations with similar features found in Earth's northern tundra, the polygons were believed to have been created by repeated freezing and thawing of ice mixed with the soil.

This artist's impression shows Phoenix making its powered descent onto the surface of the Vastitas Borealis, seconds before touchdown.

By chance, dust blown away by the retro rockets during the last stages of descent revealed brighter material underneath the surface, which analysis soon showed was rich in ice.

Other discoveries included calcium carbonate minerals (another sign of a warm, wet past), and perchlorate salts (a find that has potential implications for both the possibility of Martian life, and the prospects for future colonization).

Phoenix packed a great deal of discovery into a relatively short operational lifetime. Originally intended to be active for around 90 sols, it experienced its first sunset in early September, but continued to cling onto life for a further 50 sols. The lander made its final transmission on November 2 before contact was lost when it was overwhelmed by the increasing hostility of a Martian winter.

This panoramic view of Phoenix's landing site was compiled from hundreds of separate images taken over several weeks.

## Mars Science Laboratory

Spirit and Opportunity had only just begun their marathon explorations of Mars when NASA announced plans for another, more ambitious rover to be launched in 2009. With dimensions 50 percent larger than the previous MERs, and a mass of 1,982 lb (899 kg), this rover would form the key element of a mission designated Mars Science Laboratory.

Technical delays and an overrunning budget forced a reluctant decision in 2008 to postpone the spacecraft's launch, missing the 2009 window, instead targeting late 2011. By this time, the rover had been named Curiosity following an online poll and, more importantly, a landing site had been selected. Curiosity was to touch down at Gale Crater, a 96-mile (154-km) crater just south of the equator with a large mountain of apparently sedimentary rock at its center. The mission finally launched on November 26, 2011, aboard an Atlas V rocket. Arrival at Mars came on August 6, 2012, the spacecraft executing a series of complex, precisely timed maneuvers to put the rover on the surface.

**RIGHT** An artist's impression captures the Curiosity aeroshell as it begins its plunge through the Martian atmosphere—a journey described by engineers as "seven minutes of terror."

## High-speed Descent

After initial separation from the "cruise stage" that had brought them from Earth, the descent stage and rover plunged into the Martian atmosphere in a protective aeroshell.

Atmospheric friction slowed the incoming spacecraft from an initial speed of 3.6 miles per second (5.8 km per second) down to about 1,500 feet per second (470 m per second) in the space of four minutes, heating it up to about 3,800°F (2,100°C). At the end of this phase, roughly 6 miles (10 km) above the surface, the aeroshell deployed a parachute to slow its descent still further.

At an altitude of 2.3 miles (3.7 km), the front half of the aeroshell separated and fell away, leaving the backshell to continue its plunge through the Martian atmosphere with the descent stage and rover clamped beneath it. The Mars Descent Imager (MARDI), situated on Curiosity's underside, also began operating, snapping four high-resolution images of the approaching surface each second.

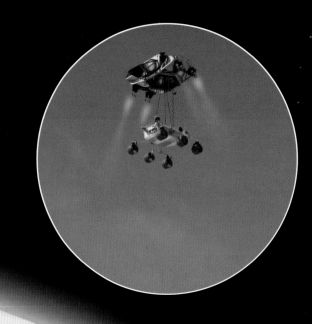

## Soft Landing

At an altitude of 1.2 miles (1.8 km) and a speed of 220 mph (100 meters per second), the backshell uncoupled and the descent stage and rover were released.

The descent stage was equipped with eight rocket thrusters that fired as the rover was lowered beneath it on a 25-foot (7.6-m) tether. Curiosity was exposed to the elements during the last stage of its descent, automatically unfolding its wheels. Finally, four of the descent stage's engines cut out, allowing it to lower the rover onto the surface before flying away and crashing at a safe distance once the tether was released.

Every stage of the landing procedure functioned according to plan, Curiosity landing within 1.5 miles (2.4 km) of its intended target, close to the central mountain of Gale Crater, after a journey of 350 million miles (563 million km).

**TOP** This graphic shows the final stages of Curiosity's descent, with the aeroshell discarded and the rover suspended precariously beneath its "skycrane."

**LEFT** Curiosity took this self-portrait over Sols 84 and 85 of its mission, in autumn 2012. It is composed of dozens of separate images from the Mars Hand Lens Imager (MAHLI) camera.

## Curiosity's Cameras

The rover carries no fewer than 17 cameras, including two full-color MastCams (with different fields of view) mounted on a pillar that raises them to 7 feet (2.2 m) above the surface, two pairs of monochrome NavCams for stereoscopic imaging of the surface, and four pairs of HazCams, mounted low on the chassis, for identifying and imaging obstacles in 3-D. In addition, the MARDI descent imager on the rover's underside can take high-resolution images of the Martian soil, and the Mars Hand Lens Imager (MAHLI), mounted on the end of Curiosity's robot arm, can capture images of everything from microscopic soil grains to the body of the rover itself.

The final camera, ChemCam, is a complex instrument mounted at the top of the rover mast and used to analyze the composition of surface rocks at a distance. It consists of a powerful infrared laser that can vaporize rock samples at a distance of several feet. Cameras then photograph the surrounding context, and analyze the light emitted from the evaporating material across the infrared, visible, and ultraviolet parts of the spectrum, revealing its elemental composition.

Halfway up the mast sit the sensors for the Rover Environment Monitoring Station (REMS), a weather station that includes, for the first time on Mars, an ultraviolet sensor for measuring the levels of dangerous UV radiation penetrating the thin atmosphere to the planet's surface.

## Warming Refrigeration

The rover's body is a substantial box filled with technology including control units for the various scientific instruments, and the probe's twin computers, each of which uses special components designed to resist damage from radiation such as cosmic rays.

The interior is filled with more than 200 feet (60 m) of internal tubing, through which refrigeration fluid is pumped in order to regulate temperatures. While the same sort of fluid is typically used in fridges on Earth, for much of the time on Mars it is actually warming the rover's components.

The entire rover moves on a scaled-up version of the rocker-bogie suspension system successfully used by Spirit and Opportunity. Its six wheels have a diameter of 20 inches (50 cm), and each is powered by an independent motor.

## Nuclear Generator

One key difference in Curiosity's design was the absence of the solar panels that had been so vital to previous rovers. The rover's size and the power demands of its experiments meant that solar power was impractical. Instead, NASA equipped the vehicle with a radioisotope thermal generator (RTG), a power unit capable of generating electricity using the heat produced by a sample of radioactive material, in this case 10 lb 7 oz (4.8 kg) of plutonium-238 dioxide. Curiosity's was the first large-scale RTG to be landed on the Red Planet since the Viking missions some 36 years previously.

Freed from reliance on the Sun, Curiosity's drivers had no need to find suitable locations for overwintering on sunny slopes. The power plant is mounted at the back of the vehicle, close to the high- and low-gain antennae it uses for communication with Earth.

Below the antennae at the rear sits the DAN (Dynamic Albedo of Neutrons) experiment, a Russian-built instrument that measures the hydrogen and ice composition of soil beneath the rover, employing similar principles to those used on the *2001 Mars Odyssey* probe.

LEFT A close-up view of Curiosity's remote sensing mast shows instruments including (from top to bottom) ChemCam and its accompanying laser, twin Mastcams, and (beneath the motorized pivots), the various elements of the Rover Environment Monitoring Station.

**ABOVE** A close-up view of the tools mounted on the hand or turret at the end of Curiosity's robot arm. This photograph, from the rover's left mastcam, was taken on Sol 174 of the mission, as Curiosity prepared to begin drilling at the John Klein rock site.

**LEFT** A close-up view of the inlet filter where Curiosity "ingests" powdered rock samples for processing in its onboard SAM and CheMin laboratories.

## Sniffing Out Minerals

A robust robot arm, emerging from the front of Curiosity, carries several tools and scientific instruments. In addition to the MAHLI camera, Curiosity's "hand" (technically known as the turret) is fitted with an Alpha-Particle X-ray Spectrometer (APXS) instrument, a chemical sniffer for analyzing the mineral composition of rocks. Alongside these two experiments are tools for collecting samples—a drill, brush, scoop, and sieve.

The arm itself has three joints, allowing for a wide range of movement, including folding back in order to deliver samples through funnel-like inlets to two automatic laboratories, called CheMin and SAM, mounted on the front of the probe's body. CheMin (Chemistry and Mineralogy) fires X-rays into a sample and records the "X-ray diffraction" patterns that reveal its crystalline structure, as well as analyzing how the material fluoresces, offering further clues to its chemistry.

SAM (Sample Analysis at Mars), meanwhile, carries out a variety of experiments to look for organic (carbon-based) molecules in the soil and also to monitor the chemistry of atmospheric gases.

## The Lie of the Land

Following an initial week of post-landing work that included upgrading its onboard software, the Curiosity team began to test the rover's systems and instruments, gathering data about its immediate surroundings.

The lander had touched down on the floor of Gale Crater on a plain known as Aeolis Palus, roughly 5 miles (8 km) from the base of the crater's huge central mountain, Aeolis Mons, which rises about 3.5 miles (5.5 km) above the crater floor and was to be the rover's main objective.

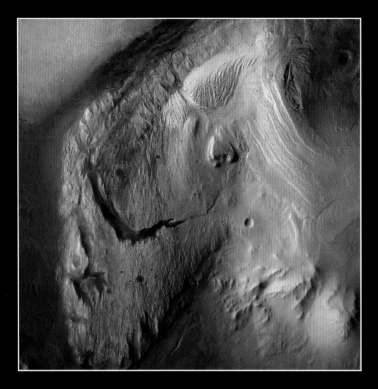

**ABOVE** A false-color image from *2001 Mars Odyssey's* THEMIS cameras reveals differences in the thermal properties of the landscape around Curiosity's landing site. Blue colors indicate relatively loose, dusty material while exposed outcrops of harder rock appear in yellow and red.

**BELOW** Sunset in Gale Crater—this spectacular panorama of Curiosity's view toward the crater's central peak combines images captured during Sols 170 and 176 of the rover's mission (in February 2013). The Sun has been reinstated during image processing to compensate for overexposure in that part of the image.

Using orbiter and ground-based photos, different routes to the mountain were identified. The most interesting potential route was selected and the rover's primary mission of two years would allow it to reach the foothills. Following the success of Spirit and Opportunity, however, scientists and engineers alike were hopeful that Curiosity would operate for much longer.

## The Glenelg Route

Curiosity's first proper drive on the Martian surface began on August 29, 2012. Its chosen route first took the rover to a region called Glenelg, southeast of the landing site, where three different types of terrain appeared to intersect. Even before it had reached this first way station, however, the mission was already making important discoveries. A pair of rocky outcrops found along the way, soon named Hottah and Link, proved to be composed of conglomerate rocks—large quantities of small, gravelly stones cemented together into a solid mass.

Conglomerates form when pebbles are transported by flowing water before being deposited to accumulate in specific locations. The shape and size of some individual pebbles, in particular their smooth, rounded appearance, indicated that this was no one-off occurrence. They had clearly been submerged for a long time, and subject to the distinctive erosion patterns that occur in water. Geologists were even able to estimate the nature of the flowing water—a stream less than 3 feet (1 m) deep, flowing at a speed of roughly 3 feet per second (1 m per second).

Satellite images confirmed the likely origin of these gravels. Curiosity was traversing an alluvial fan, a wedge-shaped deposit of sediments left behind as water flowed into Gale Crater through a channel known as Peace Vallis. The water, and the pebbles it carried within it, had probably traveled a long way across Mars to reach this point.

**ABOVE** Curiosity's landing site (green dot) and targeted landing ellipse (blue circle) are shown in context on a perspective view of Gale Crater. This image was created using elevation data from ESA's Mars Express stereo camera, high-resolution monochrome images from Mars Reconnaissance Orbiter's context camera, and color data from the Viking orbiters.

ABOVE Even Curiosity's wheel tracks can provide useful information about the Martian soil—imprints like this one at the "Rocknest" site can show how particles of different sizes are distributed within sand ripples.

## Rocknest and Beyond

In early October 2014, Curiosity's track took it past a light-colored patch of sand nicknamed "Rocknest." Here, scientists spotted a number of bright particles in the sand.

At first these were assumed to be fragments of metal discarded during the rover's descent, although when similar particles came to light elsewhere, it became clear that they are actually a natural component of the Martian soil. Some of the loose sand was collected for analysis in Curiosity's CheMin laboratory, and X-ray diffraction tests identified a number of minerals in the dust, including feldspar and olivine. Geologists back on Earth concluded that the soil was probably produced by the breakdown of volcanic rocks such as basalt.

## Water for Settlers

At the same time, another soil sample was scooped into the SAM instrument, designed to search for volatile and organic compounds. When their results were announced almost a year later, mission scientists were able to confirm the presence of a significant amount of water in the soil.

Despite the earlier "water map" of 2001 Mars Odyssey and Phoenix's confirmation of an icy permafrost at the poles, the discovery that, even at equatorial latitudes, the uppermost layer of soil contained as much as 2 percent water came as a surprise. Importantly, it suggests that future colonists on Mars should be able to extract plentiful water simply by heating and processing relatively small amounts of soil.

The SAM instrument also seemed to confirm the presence of carbonate rocks formed in damp conditions, but a less positive discovery was the confirmation of perchlorate salts similar to those found by Phoenix. The confirmation that these highly reactive chemical compounds are present in Martian soil at all latitudes may significantly affect the prospects for Martian life in both the past and present.

## Twinning Ceremony

Curiosity arrived at its first target, Glenelg, early in 2013 and began its investigation by taking detailed photographs of the area to help scientists back on Earth identify suitable targets for further study. The features in this region were named after localities in the Yellowknife area of northern Canada, but the residents of the original Glenelg village in Scotland still celebrated the Curiosity's arrival with a "twinning ceremony."

LEFT A color view of Rocknest shows its accumulation of windblown sand and dust downhill of a cluster of darker rocks. The sand was chosen for first use of the scoop on Curiosity's robot arm, and subjected to intense study that revealed unusually bright particles. At first, these were thought to be debris from the spacecraft itself, but further research showed them to be a natural component of the Martian soil.

**LEFT AND INSET** Close-up images of Rocknest, as it appears under Martian atmospheric conditions (main picture), and as it would appear under Earth-like lighting (inset). The rounded rock just above center in each image is about 8 inches (20 cm) across.

Three different types of terrain overlap In the Glenelg area. Bright underlying bedrock is covered by a darker landscape pitted with many small craters, the third terrain being the same type of surface on which the rover had originally landed, with somewhat fewer, larger craters. Based on images captured by Mars Reconnaissance Orbiter, geologists believed this part of the crater floor was an ancient lakebed. To prove this, they wanted to investigate the bedrock layer in a depression known as Yellowknife Bay. The rocks here proved to contain veins of a light-colored mineral and small grainy spherules.

The ChemCam instrument targeted its infrared laser onto a vein of the light material, detecting emissions of calcium and sulfur in the light of the vapor.

This suggested that it could be gypsum (hydrated calcium sulfate), similar to the Homestake Formation discovered by Opportunity. In some places the light-veined material had eroded to leave narrow cracks in the remaining rock, while in other areas the rock had eroded, leaving the veins behind as small ridges in the soil.

ABOVE An image of the John Klein Rock in Yellowknife Bay, centered on the spot chosen for Curiosity's first drilling.

## Drilling on Mars

In February 2013, Curiosity arrived at a rocky outcrop known as John Klein. Here it accomplished a notable first for the Red Planet, drilling a rock sample from beneath the surface for analysis. Despite carefully selecting the target, none of the Curiosity team anticipated the impressive results.

Curiosity had first spotted the unusual rock in panoramic images of Yellowknife Bay taken in late January. Selecting it as a target for drilling, the NASA team named it in honor of the mission's former deputy director, who had died in 2011.

Drilling began on February 8, excavating a small hole just over half an inch (16 mm) across and 2.5 inches (64 mm) deep. Immediately, the mission scientists could tell that the rock was unusual. The dusty powder that emerged from within the hole was gray-white, rather than the more familiar rusty brown uncovered by Opportunity's rock-grating experiments.

About a tablespoon of the powder was ground, pulverized, and sifted before being passed to the CheMin and SAM analytical laboratories. The scientists were astonished by what they found. The rock contained 20 percent phyllosilicate clay minerals that could only have formed in freshwater conditions.

## Conditions For Life?

Calcium sulfate, meanwhile, showed that the environment was probably mildly alkaline. Although no carbon-based organic compounds showed up in the results, the elements confirmed within the sample included carbon, oxygen, nitrogen, sulfur, and phosphorus—all important ingredients in the development of life as we understand it on Earth.

Tracking down evidence for habitable conditions in the Martian past had been one of Curiosity's major science goals, and it had accomplished it in style at the first attempt.

As one NASA scientist put it, "We have found a habitable environment which is so benign and supportive of life that probably if this water was around, and you had been on the planet, you would have been able to drink it."

In fact, the results were so surprising that scientists immediately planned to drill another sample for independent confirmation. This second rock, known as Cumberland, was a mere 10 feet (3 m) from John Klein, and of similar age, but showed evidence of a more complex history.

## Memory Glitch

Plans for Cumberland were put on hold in late February when Curiosity developed its first major fault—its primary computer began to reboot continuously due to a memory error. By the time engineers had switched control of the rover to the backup computer and confirmed everything was fully operational once again, a conjunction was approaching in which Mars and Earth lay on opposite sides of the Sun and communication was limited.

Throughout most of April, therefore, Curiosity was left largely to its own devices, carrying out a limited range of measurements for later transmission back to Earth, while remaining mostly stationary.

The Cumberland rock was finally drilled in late May, and subsequent analysis showed that John Klein was no fluke. These rocks had formed in a hospitable environment. Other forms of analysis helped to confirm the age of Cumberland and related rocks at between 3.86 and 4.56 billion years.

ABOVE The first drill-hole on Mars, photographed by Curiosity on Sol 182 (February 8, 2013). The pale color of the exposed rock powder came as a surprise to nearly everyone involved with the mission.

A pair of zoom images from Yellowknife Bay reveal distinctive veins and nodulelike concretions in the rock. On Earth, such structures are usually attributed to mineral formation as dissolved chemicals precipitate out of water in the rock.

## Trek to Mount Sharp

Curiosity finally departed the Glenelg area in June 2013, now heading directly for its primary target, Gale Crater's central peak of Aeolis Mons, informally named Mount Sharp after the geologist and Mars expert Robert P. Sharp. The drive across more than 5 miles (8 km) of varied terrain, was expected to take from nine months to a year, factoring in stops along the way to investigate interesting features.

By early August, as it celebrated its first birthday on the Red Planet, the rover had already driven more than 1 mile (1.6 km) on the surface, and sent back more than 70,000 images. The CheMin laser, meanwhile, had been used to analyze some 2,000 individual samples.

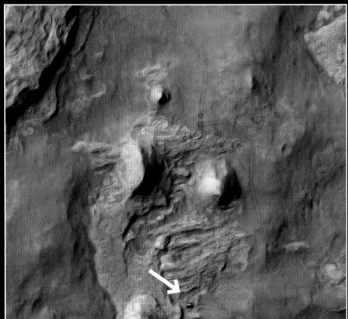

**LEFT** An overhead view of Curiosity (arrowed) amid surrounding rocks, captured by Mars Reconnaissance Orbiter in April 2014.

In May 2014, Curiosity discovered this large 7-foot (2-meter) iron meteorite, nicknamed "Lebanon." This image combines a wide-angle view of the meteorite from MastCam with details from ChemCam, and reveals various features that may be linked to erosion after the meteorite's arrival on Mars.

As it traveled, the rover took atmospheric measurements and, back on Earth, scientists reaped the rewards. Significant results from the SAM instrument included confirmation of the levels of two argon isotopes in the Martian atmosphere.

It transpired that, compared to planets elsewhere in the solar system, including Earth, Mars has a significant surplus of the heavy form argon-38 over the lighter form argon-36. Since in the early days of the solar system, the ratio must have started out the same as on other planets, the current imbalance is a confirmation that the young Mars lost much of its atmosphere (including quantities of the lighter argon-36) into space. This discovery provided a signature measurement to help confirm that some meteorites found on Earth really do originate on Mars.

Another significant atmospheric measurement was the apparently negative result for methane. Previously suggested as present in small quantities by Earth-based and orbiting instruments, methane would have major implications for our understanding of Mars, and its absence was a disappointment. Yet this was not the whole story…

In September 2013, Curiosity arrived at an outcrop of light-streaked rock called Darwin. The rover paused for four days, studying rock that proved to be a water-formed conglomerate shot through with veins of light material. Though not the gypsum found at Yellowknife, these veins indicated that the rocks of Darwin had water running through them at some point after their formation.

## Into the Foothills

In late October 2013, Curiosity paused for a couple of days to study a second waypoint called Cooperstown, but the rover's progress was temporarily halted in mid-November by the detection of a potential short circuit. Fortunately, this was eventually traced to the rover's RTG radioisotope power supply, and found to be harmless.

Shortly after getting underway again, engineers transmitted a new upgrade to Curiosity's operating software (the third since landing). This improved the rover's speed still further by making it easier to program two-day drives. However, the MAHLI camera on Curiosity's robot arm was also used to carry out a photographic inspection of each of the rover's wheels and assess how they were coping after a recent journey across sharp rocky terrain.

After assessing the likely risk of continued wear and tear on Curiosity's wheels, February 2014 saw engineers take action to avoid further damage: they sent Curiosity on a short-cut over a small sand dune to avoid more treacherous ground. They also tested driving in reverse as another possible way to reduce damage, and generally took more care in selecting terrain to drive through, even though this slowed down the rover's progress.

**BELOW** An uphill view toward the Pahrump Hills from Curiosity's Mastcam, captured on Sol 751 in September 2014. Colors in this image are balanced to mimic Earthlike lighting conditions.

March finally brought the next waypoint within view, a site called the Kimberley. This complex outcrop comprises multilayered sandstones whose physical differences suggest they were laid down at different times and in different conditions.

## The Kimberley Intersection

Four different rock units intersect in the Kimberley, and mission scientists hoped to learn how the grains within each are held together to create rocks with different resistance to erosion. At the end of April, Curiosity began drilling into a flat sandstone slab named Windjana. Subsequent analysis revealed that it was rich in magnetite, an iron compound that forms on Earth through the action of water on volcanic basalt.

The slab also contained a more varied range of clay minerals than those found at Yellowknife Bay, and a complex mineral called orthoclase (confirmed for the first time on Mars), whose presence may indicate rock that went through repeated incidents of melting. Since the sandstones at the Kimberley are thought to derive from materials transported down from the rim of Gale Crater, these findings reveal hidden secrets about the surrounding regional geology.

Setting off again in May, by late June Curiosity had crossed beyond the relatively smooth terrain that defined its original target landing zone and was confronted with a hazardous region of sharp rocks known as Zabriskie Plateau. There was no practical detour, so the rover spent much of July picking its way through the boulder field. Its wheels emerged mostly

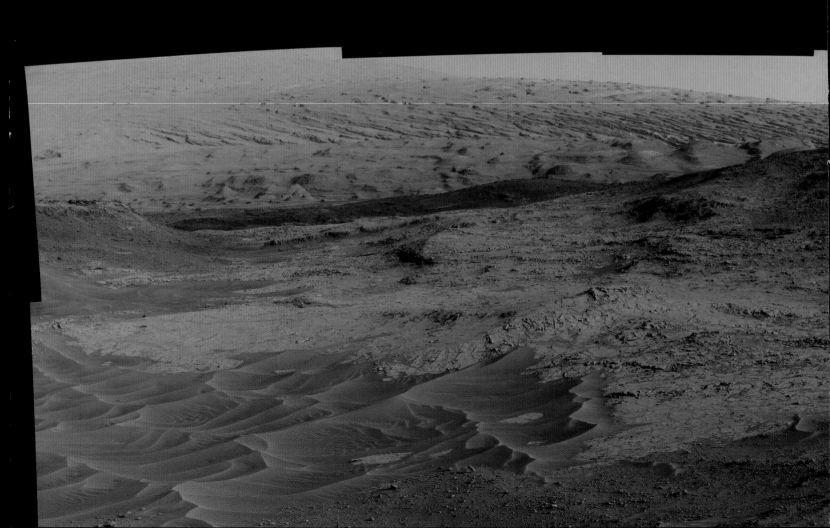

unscathed, and it was soon back on track, heading for the first outcrop of Mount Sharp itself—Pahrump Hills.

In August, problems crossing the sand-filled Hidden Valley forced the rover to back up. Unable to reach the Pahrump area, the team pinpointed a flat rock that appeared to have a similar origin to Pahrump, and looked quite different from the

**ABOVE** A NavCam image shows the ramp by which Curiosity entered and later retreated from the treacherous sand-trap.

sandstones previously encountered. Plans to drill this rock, named Bonanza King, were abandoned after initial tests found that it shifted under light pressure from the drill. Instead, Curiosity pressed on for the foot of Mount Sharp itself.

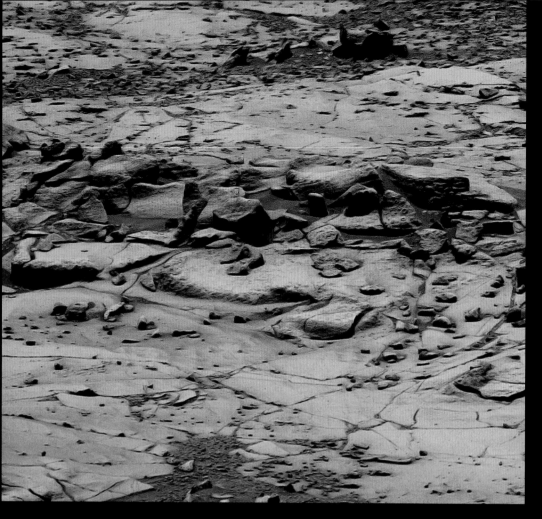

LEFT Detail from "Pink Cliffs," a rocky ridge within the Pahrump Hills. The ridge's apparent resistance to erosion compared to its surroundings attracted the attention of NASA scientists, and it was here that they identified the drilling target known as Mojave.

BELOW Multiple layers of sandstone around the Kimberley area, roughly 1 mile (1.6 km) north of Mount Sharp, are now thought to have formed when sediments waere deposited from a broad river that flowed into Gale Crater and slowed rapidly as it encountered the standing water of the crater lake.

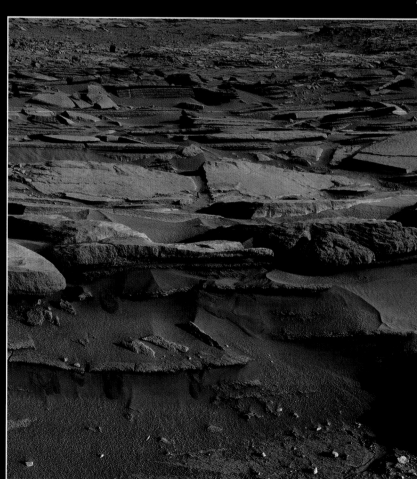

# Climbing Mount Sharp

The Curiosity team now decided to change the rover's route of approach. Initially they had planned to investigate a group of low mounds called the Murray Buttes before crossing onto the mountain itself, but after concluding that they were probably similar to Bonanza King, they now took a short cut onto the mountain itself, passing through an entry point close to the Pahrump Hills in early September 2014.

Access to the mountain slope involved passing south along a sandy-floored valley. Fortunately this proved to be far less treacherous than the Hidden Valley where the rover had been forced back a few weeks before. By late September, Curiosity was able to take its first drill sample from the base layers of the mountain at the backside of the Pahrump Hills.

In stark contrast to the gray powder revealed in Yellowknife Bay and the light-colored dust at Windjana, this sample proved to be as red as the mountain's surface. CheMin soon showed that it was rich in the red iron ore hematite, confirming observations from Mars Reconnaissance Orbiter.

The presence of magnetite and other minerals found elsewhere indicated that minerals with different "oxidation states" were able to co-exist when the rock formed—an important discovery, since microbial life on Earth often generates energy by transforming chemicals from one oxidation state to another.

With an initial sample of Mount Sharp in hand, the mission now followed a predetermined strategy: retracing the route to take spectroscopic measurements of target rocks with APXS and microphotography with MAHLI, in order to specify which rocks would be drilled by Curiosity on a third pass.

Meanwhile, back on Earth, the Curiosity team announced two major discoveries related to the possibility of Martian life—a sudden spike in levels of atmospheric methane detected by SAM around the end of 2013, and SAM's detection of organic (carbon-based) chemical compounds in the Cumberland rock sample. Such compounds are not evidence for ancient life in themselves, but they are the building blocks of all life.

By mid-January 2015, the scientists had settled on Curiosity's next drilling target—a light-colored rock called Mojave that contained large numbers of slender features about the size of rice grains. The rock broke apart during an initial drilling test, so, after a week's pause for another software upgrade, the rover moved on to a similar nearby rock, Mojave 2. It used a new, low-impact technique to extract powder from the relatively fragile rock. Initial analysis of the sample suggested the rock had formed in more acidic conditions than those found among the crater floor rocks.

BELOW The distinctively tilted patterns seen in this rock outcrop, known as cross-bedding, are the result of waves disturbing loose sediment on a shallow lake floor within Gale Crater.

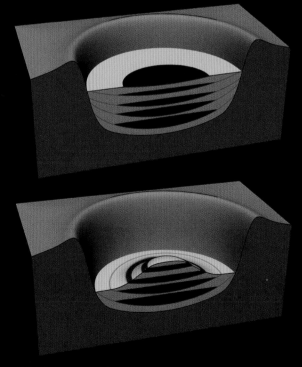

ABOVE A model of Gale Crater's geology. It seems that the crater was once filled with sediment layers from the surrounding highlands, with deeper layers deposited in a central lake that disappeared and reappeared over a long timescale (top). The softer surrounding sediments have since eroded more easily, leaving the harder rock of Mount Sharp exposed in the center of the crater (bottom).

# ONWARD TO MARS

Thanks to the ongoing armada of space probes both orbiting Mars and landing on its surface, we have learned more about the Red Planet since the dawn of the 21st century than we did in all of previous human history. Discoveries from Mars Express, Mars Reconnaissance Orbiter, the Mars Exploration Rovers, and Curiosity have transformed our view of the planet yet again, painting a picture of a world with a hospitable past and a not-so-hostile present. So the obvious question is: What next?

## Rocks from Mars

Each new launch window sees other probes joining the Martian flotilla—now frequently aiming to answer specific questions rather than simply conducting reconnaissance. And many more unmanned missions are either in advanced planning stages or actively under construction.

One medium-term goal of NASA's program is a sample return mission that can land on Mars, collect rocks and dust, and bring them back for analysis by the full range of scientific tools available in a modern laboratory. Such a mission would be fraught with difficulties, not least in ensuring that the precious sample remained uncontaminated by materials from Earth, and is likely still to be a decade away from a potential launch.

In the meantime, however, geologists are making do with the next best thing in the form of Martian meteorites—rocks whose link to the Red Planet has long been suspected, and whose origin has recently been confirmed by measurements from the Curiosity rover.

These meteorites have proved an abundant source both of information and controversy, thanks to daring claims that they may show evidence of ancient Martian life.

## Settlements in Space

The quest for signs of Martian life remains a key priority for NASA and other space agencies. In January 2014, faced with abundant evidence for habitable climates on ancient Mars, NASA re-dedicated the Opportunity and Curiosity rovers specifically to this search as a primary goal. Future missions will carry specialized experiments to detect past and present life, following the legacy of the 1970s Viking experiments and the ill-fated Beagle 2 probe.

For some, continued unmanned exploration is unlikely to answer the key questions. They argue that space agencies and private enterprises should be spearheading efforts to land humans on Mars as soon as possible. Plans range from the ambitious MarsOne project through to the ingenious Mars Direct scheme and NASA's scientifically rigorous approach.

Putting people on Mars involves fearsome challenges that will push engineering and human endurance to its limits. When such a mission eventually happens, it will need to combine the best ideas, people, and technology from both public and private sectors, and it will probably mark the beginning of a whole new phase in human history—the settlement of our solar system.

**ABOVE** As early as 1963, NASA held a symposium on the possibility of manned planetary missions to follow on from the Apollo Moon program. This artist's impression depicts a proposed Mars lander spacecraft.

**RIGHT** A nuclear thermal propulsion "transfer vehicle" prepares to rendezvous with astronauts returning from the Martian surface 120 miles (200 km) above Tharsis in this artist's concept. Nuclear thermal propulsion has the potential to offer twice the efficiency of a chemical rocket, increasing either the size of payloads or the speeds with which they could transfer from Earth to Martian orbit.

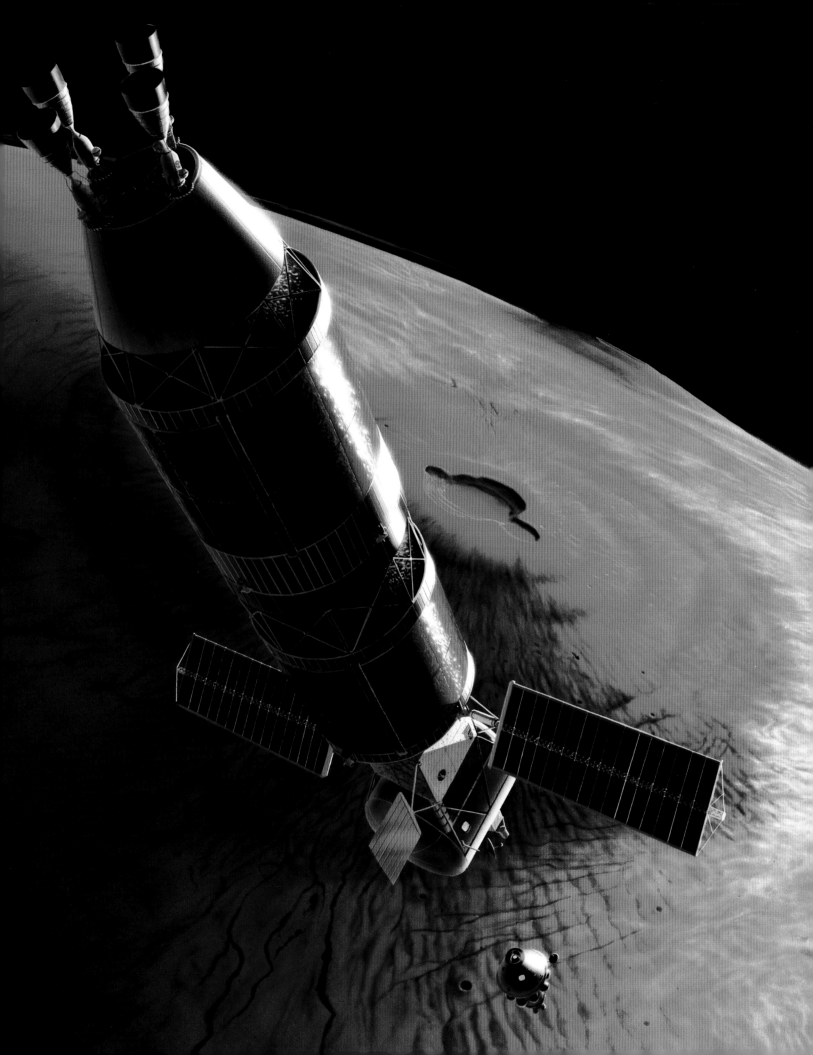

## Meteorites from Mars

Closer to home, recent years have seen geologists develop a promising new way of studying the Red Planet. Since the early 19th century, scientists have recognized that Earth is regularly bombarded by rocks from space, and attempted to classify these meteorites into different types. Most can be explained as either primordial material from the early days of the solar system, or the shattered remnants of asteroids that formed out of this material.

A few exceptions seem to be fragments of objects with a far more complex past. By the early 1980s, three specific groups, the shergottites, nakhlites, and chassignites (each named for their "prototype" meteorite, the first of their kind to be classified), had been identified as potential fragments of Mars. They appeared to have much younger origins, contained traces of chemical weathering in the presence of water, and bore some similarities to Martian surface rocks analyzed by the Viking landers. In 1983, scientists succeeded in extracting and analyzing gas bubbles trapped by glass inside a shergottite, and found further similarities to the Martian atmosphere.

Today, more than 130 meteorites from Mars have been identified, and as the closest thing we have to a sample of Martian crust for analysis in Earth's laboratories, they are the subject of intense interest. But teasing out the secrets of their Martian origin is complicated due to their well-traveled history.

**ABOVE** Shergottites such as NWA 6963 are a group of Martian meteorites with volcanic origin.

**ABOVE** Tissint meteorites come from an igneous Martian rock that fell to Earth in Morocco in 2011.

The raw materials of each meteorite originated as rock created at some point in the planet's long history, and after an indeterminate length of time, subsequently shocked and transformed by whatever event ejects them into space (most likely a large meteorite impact). This kind of event can fundamentally alter the structure of the rock itself, although it may also create glassy areas within the rock that can trap samples of the Martian atmosphere or other volatiles.

## Forces of Change

The meteorite then spends another long, indeterminate spell orbiting in interplanetary space, before eventually colliding with Earth. A fiery entry into our planet's atmosphere can leave its outer layers charred and transformed, and if it subsequently spends some time on the ground before its true nature is recognized, it can be further modified by the influence of Earth's atmosphere and water.

Consider, for example, the most famous Martian meteorite, ALH 84001. This lump of rock weighing 4 lb 3 oz (1.9 kg) was discovered in the Allen Hills of Antarctica in 1984—prime meteorite-hunting territory. Any rock sitting on or within the upper ice layers stands out amid its pure white surroundings, and can only have got there by dropping down through the atmosphere from outer space.

ALH 84001 is thought to have formed from volcanic activity some 4.5 billion years ago, and features within it suggest transformation by a powerful shockwave, most likely from a nearby impact, about 500 million years later. At some point in the following billion years, globules of carbonate minerals were deposited in the rock, perhaps by the action of flowing water.

After a long period on or beneath the surface, it was ejected into space around 16 million years ago, and plunged to Earth some 13,000 years ago. The latter two dates are gleaned from measurements of the rate at which cosmic rays striking the meteorite in space enriched it with certain radioactive isotopes, and the rate at which these have subsequently declined since the meteorite's landing in Antarctica.

**BELOW** A panoramic Martian landscape captured by Curiosity on Sol 538 of its mission (February 9, 2014). By analyzing the crystals within igneous Martian meteorites, scientists can estimate the date when they formed, revealing a wide range of ages from several billion to just a couple of hundred million years.

**ABOVE** Landscapes such as the Nubian Desert and the wastes of Antarctica are ideal territory for meteorite-hunting, since rocks from space stand out as unusual amid the surrounding terrain.

Hot spring environments on Earth, such as the Grand Prismatic Spring of Yellowstone, Wyoming, are home to flourishing microbial life—could similar springs on Mars have given rise to past Martian life?

## Signs of Life?

ALH 84001 is the oldest known Martian asteroid by some 3 billion years, and hence the only one to give us an insight back into the period of Martian history when life might perhaps have found a foothold. Little wonder, perhaps, that it has been subjected to intense studies ever since its discovery. Even so, few would have anticipated the astonishing claims put forward for this unassuming lump of rock by a team of NASA scientists in 1996.

**ABOVE** The controversial 4-lb (1.9-kg) meteorite ALH 84001, said by some NASA scientists to contain traces of ancient Martian microbes.

The team, led by David McKay of NASA's Johnson Space Center, reported finding traces of carbonate minerals, the iron compound magnetite ($Fe_3O_4$), and even tiny, wormlike structures embedded deep within the meteorite.

The carbonates were controversial in their own right, as some of the first evidence for the warmer, wetter Martian past we now take for granted, but it was the other features that sparked a firestorm of publicity and even led President Bill Clinton to make a statement hailing the discovery, for McKay and his colleagues claimed that they were "biogenic" in origin—the product of once-living Martian micro-organisms.

Needless to say, such potentially historic claims met with scepticism in some parts of the scientific community. Some argued that, while tiny magnetite "nano-crystals" of the type found in the rock are commonly made through organic processes by Earth microbes, there are various other means of making similar molecules through geological, rather than biological means.

Others pointed out that the wormlike structures, put forward as potential fossil remains, are far smaller than any bacteria so far identified on Earth, voicing their doubts that such tiny structures could support the complex biochemistry of a living, self-replicating organism. Some even suggested that the "worms" were an artifact of the way the rock samples had been prepared for electron microscopy.

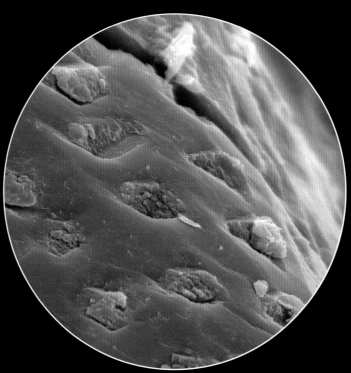

ABOVE The Nakhla meteorite, which fell to Earth over Egypt in June 1911, bears traces of carbonate and hydrate minerals, suggesting that it was subject to the actions of water at some point in its history. This scanning electron micrograph (SEM) image reveals pores and veins where carbonate are concentrated in the rock—a type of microstructure that, on Earth, is usually linked to the action of bacteria.

## Rumbling Debate

The ALH 84001 debate has now rumbled on across two decades, and seems no closer to reaching a conclusion. Other NASA teams have shown how the wormlike features can be produced by non-living means, but McKay's team have expressed doubts about such experiments beginning with unrealistically pure laboratory conditions. Others in the original group have focused on the potential alternative pathways for laying down magnetite, and shown that none can replicate the purity of the Martian crystals.

Further studies have tried to establish the conditions in which the meteorite formed. In 2005 Vicky Hamilton of the University of Hawaii compared the mineralogy of ALH 84001 with surface measurements from *2001 Mars Odyssey*, finding a potential link to Eos Chasma, in the Valles Marineris canyon system. In 2011, analysis of the carbonate nodules showed that they formed at a hospitable temperature of around 64°F (18°C) within a slowly evaporating body of water.

Microbial life thrives in far more hostile conditions than this on Earth, so is it possible that ALH 84001 does indeed hold our first proper clue to the existence of life on Mars? The evidence remains inconclusive, and it now seems unlikely that the matter will be resolved without further samples of ancient Martian rock—either dropped conveniently from the sky, or brought back by robot probes and future astronauts.

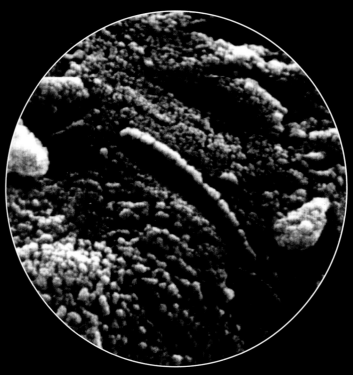

ABOVE This SEM shows wormlike structures within ALH 84001 that some have speculated might be the fossilized remains of actual Martian microbes. With diameters measured in tens of nanometers, they are far smaller than any known Earth bacterium.

## Probing the Future

While scientists hope that orbiters such as *2001 Mars Odyssey*, Mars Express, and the Mars Reconnaissance Orbiter still have a long and productive life ahead of them, it's inevitable that they will fail sooner or later. A new generation of spacecraft is being developed to supplement, and ultimately replace them.

The late 2013 launch window saw two such missions depart from Earth—NASA's MAVEN and the Indian Space Research Organization's (ISRO) Mars Orbiter Mission.

MAVEN (the Mars Atmosphere and Volatile Evolution) is the first spacecraft dedicated to studying the history of the Martian atmosphere and the effects of the solar wind, the steady stream of particles emitted by the Sun. It aims to discover how atmospheric gases and low-boiling-point "volatiles" have slipped from Mars' tenuous grasp, and the rate at which they are still disappearing today. Following a highly elliptical orbit, MAVEN makes brief, high-speed "dips" to an altitude of around 90 miles (150 km), skimming the upper atmosphere to analyze the mix of gases it contains.

Key measurements involve studying the ratio of various isotopes (atoms with the same chemistry but different weights). Simple physical processes causing lighter isotopes to escape into space or absorb heavier ones at the surface can reveal details about how the Martian atmosphere evolved, and why it is now so much thinner than it once was.

India's Mars Orbiter Mission, also known as Mangalyaan, is the country's first interplanetary probe and was designed largely as a test of ISRO's ability to launch such missions. The 1,100-lb (500-kg) spacecraft carries some 33 lb (15 kg) of scientific instruments, including an experiment that uses different techniques from MAVEN's to measure isotope ratios of hydrogen in the atmosphere, and another designed specifically to sniff out atmospheric methane.

India plans to follow up with a second Mars mission to launch around 2018 or 2020. Capable of carrying a heavier scientific payload, current plans envisage Mangalyaan-2 as a lander/rover combination.

## International Efforts

Other nations, too, have their eyes set on Mars. After the failed launch of Phobos-Grunt and Yinghuo-1, a joint Sino-Russian mission involving a Chinese Mars orbiter and a Russian sample-return from the Martian moon Phobos, China now plans to send a complex mission involving an orbiter, lander, and rover before 2020. Russia is considering a Mars sample-return mission to be launched perhaps a decade from now, and the United Arab Emirates has announced its own ambition to send an orbiter to Mars in 2021.

Meanwhile NASA's immediate plans involve InSight (Interior Exploration using Seismic Investigations, Geodesy, and Heat Transport), to be launched in 2016. This lander will study the internal structure of Mars using

**LEFT** An automated Mars Sample Return (MSR) mission is an obvious next step in the scientific exploration of Mars, and three separate mission concepts are being developed—a joint effort from NASA and ESA, and individual projects from the Russian and Chinese space agencies. This artist's concept shows a static solar-powered lander launching a cache of Martian soil aboard a small rocket. Unable to escape Martian gravity on its own, the cache would instead be picked up by a return vehicle waiting in orbit above Mars, for eventual return to Earth and laboratory analysis.

a design based on the successful Phoenix lander of 2008, and targeting an equatorial region, probably on Elysium Planitia. Its array of instruments will include seismometers to map the internal layers through their effect on seismic waves triggered by earthquakes, and a heat probe that will burrow down 16 feet (5 m) below the surface in order to establish the warmth of the Martian crust and the rate of heat escape from the planet's core.

Beyond this, the U.S. space agency is developing a rover mission to be launched in 2020. Building on the success of Curiosity, this robot vehicle will carry a suite of experiments to further investigate the potential for Martian life.

**ABOVE** India's groundbreaking Mars Orbiter Mission (MOM, or Mangalyaan) reached orbit around Mars in September 2014, making the Indian Space Research Organization (ISRO) the fourth space agency to reach the Red Planet. Largely intended as a technology demonstrator, MOM's highly elliptical orbit also allows it to send back useful images of the entire Martian globe.

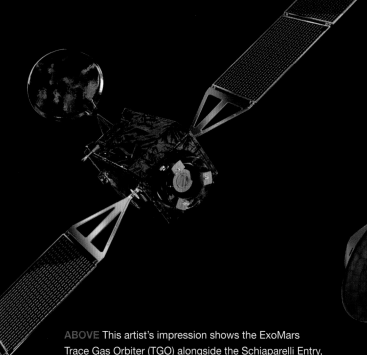

# ExoMars

Perhaps the most exciting project planned for the near term is ExoMars. A collaboration of the European Space Agency (ESA) and the Russian Roscosmos, the project is an ambitious multi-mission program that will use a variety of techniques to search for signs of Martian life (so-called "biosignatures"), both past and present.

Originally conceived in December 2005 as an ESA-led rover project to be launched on a Russian Soyuz rocket, ExoMars has had a long and complex history. A partnership with NASA saw it grow in scope to a multi-vehicle mission to be launched by NASA Atlas rockets, but when the U.S. agency pulled out due to a budget crisis in early 2012, ESA forged a new deal with Roscosmos as a full partner.

In its current configuration, ExoMars will have two distinct phases. The Mars opposition of early 2016 will see the launch of the ExoMars Trace Gas Orbiter (TGO), a project designed to map the locations of the methane outbursts that could be linked to the presence of microbial life. TGO will also act as a carrier for an "Entry, Descent and Landing Demonstrator Module" (EDM) called Schiaparelli.

Once deployed from the orbiter, the EDM's combination of parachutes and advanced Russian-built retro rockets will guide it to a safe landing. One unique innovation is a deliberately crushable "undercarriage" designed to absorb the final impact of touchdown. The primary goal is to fine-tune landing procedure for phase two of the project, but Schiaparelli also carries a limited suite of scientific instruments

**ABOVE** This artist's impression shows the ExoMars Trace Gas Orbiter (TGO) alongside the Schiaparelli Entry, Descent, and Landing Demonstrator Module, both currently scheduled to reach Mars in late 2016.

**BELOW** ESA scientists are still investigating landing sites for the ExoMars rover—this image from the HiRISE camera on Mars Reconnaissance Orbiter shows a broad swathe of one candidate site—Mawrth Vallis at a latitude of about 22°N. This ancient river valley is thought to be one of the oldest on Mars, and is teeming with light-colored rocks that are rich in a variety of clay minerals. The clays may mark an area where water and volcanic activity once interacted—a recipe for potential microbial life.

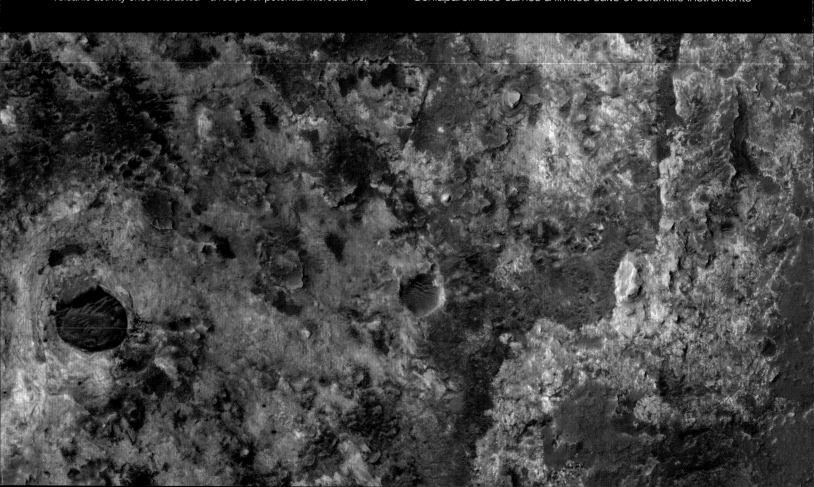

A primary goal of TGO is to identify methane hotspots that can be targeted by the ExoMars rover, a versatile robotic laboratory slated for launch at the following opposition in late spring 2018. The vehicle will arrive on a scaled-up version of the Schiaparelli EDM lander, and a ramp will then descend to allow the rover on to the Martian surface.

Roughly the size of NASA's Spirit and Opportunity, the ExoMars rover will be solar powered, although the Russian-built lander module will carry a radioisotope generator to power its on-board weather station and other instruments. Its mission is planned to last six months, with TGO acting as a relay station for sending data back to Earth. The rover's instruments will include an array of cameras, along with various devices for identifying potential targets for study. These include ADRON, a neutron spectrometer similar to Curiosity's DAN experiment, for identifying water pockets in the soil below the rover.

The rover will also carry an infrared spectrometer called ISEM, operating on similar principles to Mars Exploration Rovers' Mini-TES, and used to identify the bulk mineral properties of rocks at a distance, and a ground-penetrating radar. Once potentially interesting targets are located, ExoMars can

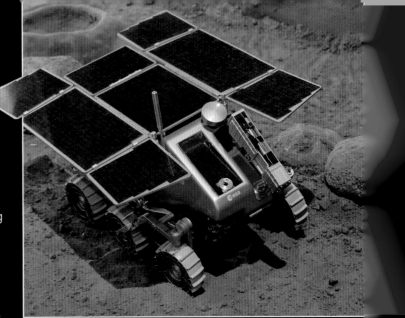

**ABOVE** The exact design of the ExoMars rover is still being determined—this model was one prototype displayed at the Paris Air Show as early as 2007.

could at least carry all the supplies they needed for a few days on the lunar surface. They also enjoyed more or less constant communication with mission control, with barely a second's delay to radio signals in either direction.

Getting to Mars, therefore, represents a huge step up in danger and difficulty. Even at the closest oppositions, the Red Planet is 35 million miles (56 million km) away. That's about 140 times farther than the Moon. Even if a bulky manned spacecraft could be boosted to the same speed as an unmanned probe, it would still take six months or more to reach Martian orbit.

More importantly, following arrival on Mars, the changing positions of the planets pretty much enforce a prolonged stay, unless a mission departs again mere weeks after its arrival. It is simply more practical to remain on Mars until the next opposition two years later. A mission will therefore take three to four years from Earth departure until its safe return, demanding a very different approach from previous manned spaceflights.

## Hazardous Mission

Based on research aboard the International Space Station, the prolonged weightlessness during most of the flight should not be an insurmountable challenge in itself: Astronauts have now spent more than a year at a time enduring similar conditions in orbit, and it seems that with diet supplements, medicines, and regular exercise, it is possible to return to normal gravity with no long-term ill-effects.

Perhaps the fact that astronauts on the first Mars missions would have to readjust to the (admittedly weak) Martian gravity without the safe environment, support staff, and recovery time available to returning space station crews actually poses a greater problem.

The most challenging element of the interplanetary flight, however, is the danger caused by radiation, including cosmic rays (actually high-energy particles) from the Sun and other sources. Earth's magnetic field and thick atmosphere combine to protect life on Earth from these damaging rays. Our planet's magnetosphere also extends well beyond the region in which most manned spacecraft orbit, shielding astronauts, too.

A long-duration interplanetary spaceflight could be very different, potentially subjecting its crew to prolonged exposure to harmful levels of radiation.

The weak magnetosphere and thin atmosphere on Mars would offer far less protection compared to Earth, even after the astronauts landed and set up base. For this reason, measuring the flux of cosmic rays on the planet's surface, and studying how the upper atmosphere interacts with particles from the Sun have been key science goals for recent Mars landers and missions such as MAVEN.

ABOVE NASA's Space Launch System (SLS) is intended to be the most powerful rocket system ever developed, capable of opening up a new era in manned space exploration. Developed in a series of phases of "blocks," the SLS should carry NASA's new *Orion* spacecraft into Earth orbit and beyond, with early targets for exploration including a near-Earth asteroid before an eventual mission to Mars. The initial "Block 1" configuration is intended for first launch in 2018.

## Routes to Mars

However sophisticated our robotic tools become, they can only go so far—and it seems inevitable that, at some point in the next couple of decades, humans will finally make the long journey to the Red Planet themselves. The scientific bounty from such a mission will be tremendous, and the technological spinoffs for those on Earth are likely to be equally influential, but the likely establishment of the first permanent human settlement beyond Earth will surely be the most important step of all.

So far, the farthest manned space missions have traveled no farther than Earth's own Moon, a mere 250,000 miles (400,000 km) away. For the Apollo astronauts of the late 1960s and early 1970s, the journey took three days in either direction, and while they were on their own if something went wrong (as indeed it did on the famous Apollo 13 mission), they

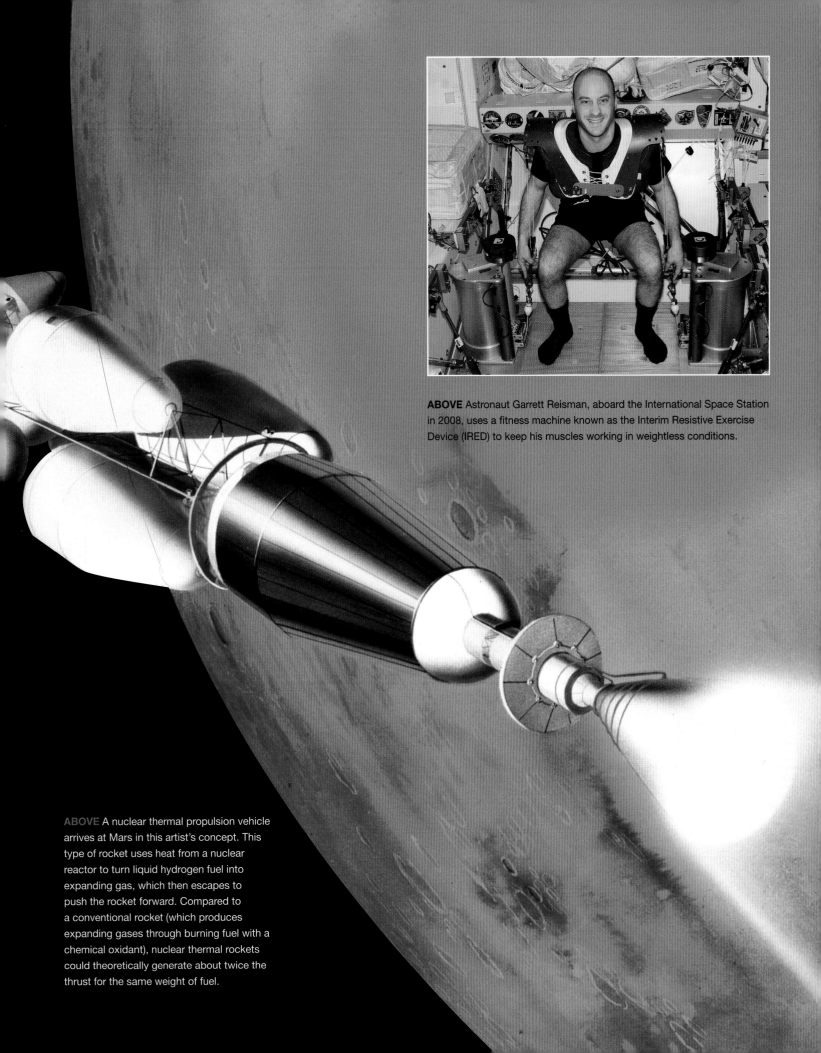

**ABOVE** Astronaut Garrett Reisman, aboard the International Space Station in 2008, uses a fitness machine known as the Interim Resistive Exercise Device (IRED) to keep his muscles working in weightless conditions.

ABOVE A nuclear thermal propulsion vehicle arrives at Mars in this artist's concept. This type of rocket uses heat from a nuclear reactor to turn liquid hydrogen fuel into expanding gas, which then escapes to push the rocket forward. Compared to a conventional rocket (which produces expanding gases through burning fuel with a chemical oxidant), nuclear thermal rockets could theoretically generate about twice the thrust for the same weight of fuel.

# Hardware for Mars

Traveling to Mars will require substantial improvements on current technology. Without such major technological changes, the journey will call for bigger chemical rockets and a larger manned space capsule capable of sustaining a crew throughout the six-month long flight.

So far, much of the hardware for proposed manned missions by the European Space Agency, Russia, China, and various commercial space companies, remains sketchy, but a few potential Mars spacecraft are at least in the early stages of design and development.

NASA's *Orion* Multi-Purpose Crew Vehicle is being developed specifically for exploration beyond the confines of Earth orbit. Its blunt, conical design harks back to the spacecraft of the Apollo era, and when coupled with a "Deep Space Habitat" module, it has the potential to sustain crews of four to six on missions lasting several months or more.

*Orion* is, however, still in the fairly early stages of its development and is not expected to carry astronauts until at least 2021, while the Deep Space Habitat remains at the concept stage. Some other proposed Mars missions (particularly those promoted by private individuals and foundations) envisage using a future variant of the SpaceX corporation's Dragon vehicle, so far the most advanced and successful spacecraft to have been privately developed, though it has yet to be rated for human spaceflight.

Any Mars-bound spacecraft would have to incorporate a range of features to reduce the health risks to astronauts on board. For instance, setting the entire spacecraft spinning throughout its journey could generate weak artificial gravity through so-called "centrifugal force," and storing liquid hydrogen fuel or water supplies in its walls could offer significant shielding against cosmic rays.

Just as important as the design of the main crewed vehicle is the availability of suitable launch vehicles for sending a spacecraft on its way to Mars. Most outline plans foresee the assembly of one or more large spacecraft in Earth orbit using multiple launches from the ground, and several vehicles are in development that might fulfil the role of a "heavy lift" launcher. These include NASA's own Space Launch System (SLS), designed to be built in various configurations depending on mission requirements. Although it uses elements of Space Shuttle technology, notably the solid rocket boosters, the SLS is not designed to be reusable but is an expendable launch vehicle more akin to Apollo's Saturn V. The initial SLS will be capable of lifting a payload of 155,000 lb (70,000 kg) with later variants intended to carry almost twice that capacity, making it the most powerful heavy lift vehicle ever, a record previously held by the Saturn V.

BELOW This artist's concept shows NASA's *Orion* Multi-Purpose Crew Vehicle above Earth at the start of a deep-space mission beyond low Earth orbit. *Orion*, the cone-shaped spacecraft in the foreground, is shown here coupled with a cylindrical service module (derived from ESA's space-station "ferry," the Automated Transfer Vehicle). In any long-duration mission, the service module would provide propulsion, power, and additional life-support supplies.

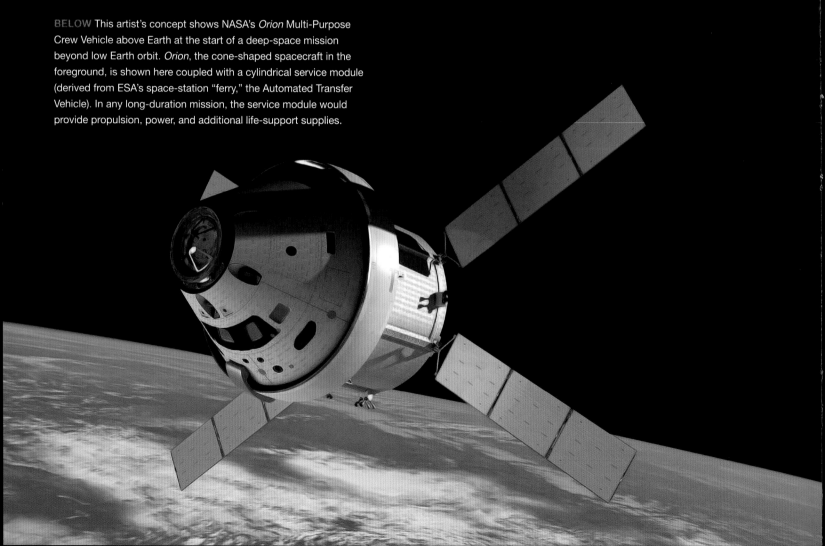

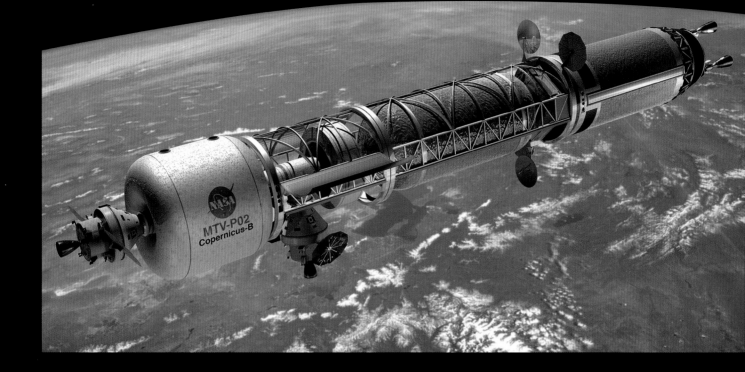

ABOVE Two *Orion* MPCV spacecraft are shown here docked with Copernicus, a nuclear-thermal transfer vehicle concept, at the start of their voyage to Mars.

SLS, having borrowed heavily from NASA's canceled Constellation program (shelved in 2010), is the design closest to completion, but there are a number of commercial rivals including SpaceX's *Falcon XX*, which is intended to have reusable rocket boosters that will land safely back on Earth.

SpaceX also has plans for the hugely ambitious MCT (Mars Colonial Transporter) which they say will be capable of ferrying 100 people to the Red Planet. Sadly, all these designs are currently still on the drawing board, and even NASA's SLS, though versatile, would prove hugely expensive to develop for a return mission to Mars.

## The Return Journey
It seems likely, then, that any future heavy-lift rocket will involve collaboration between a number of different parties, although some have argued that a large new rocket is simply unnecessary, and that mission hardware could be sent into orbit more economically on existing launchers. The spacecraft that would take its crew to Mars would then be assembled in outer space.

For most planned expeditions, sending a mission to Mars is only half the struggle—it then has to return to Earth. The enormous amounts of fuel required to lift a manned space craft into Earth orbit make sending a vehicle ready-fueled for a relaunch from Mars and return journey to Earth more or less unthinkable. In fact, despite the weaker Martian gravity, the challenge it presents is so great that in 2010, U.S. President Barack Obama instead mandated NASA simply to target a manned mission to orbit Mars by the 2030s.

Some earlier landing schemes envisaged sending the return vehicle ahead to Mars orbit, where it would wait for the astronauts to link up with it after their exploration of the Martian surface. While this solution has its advantages, recent discoveries suggest a promising alternative.

ABOVE An artist's concept of pressurized rovers and habitation modules on the surface of Mars, from a 2009 NASA study.

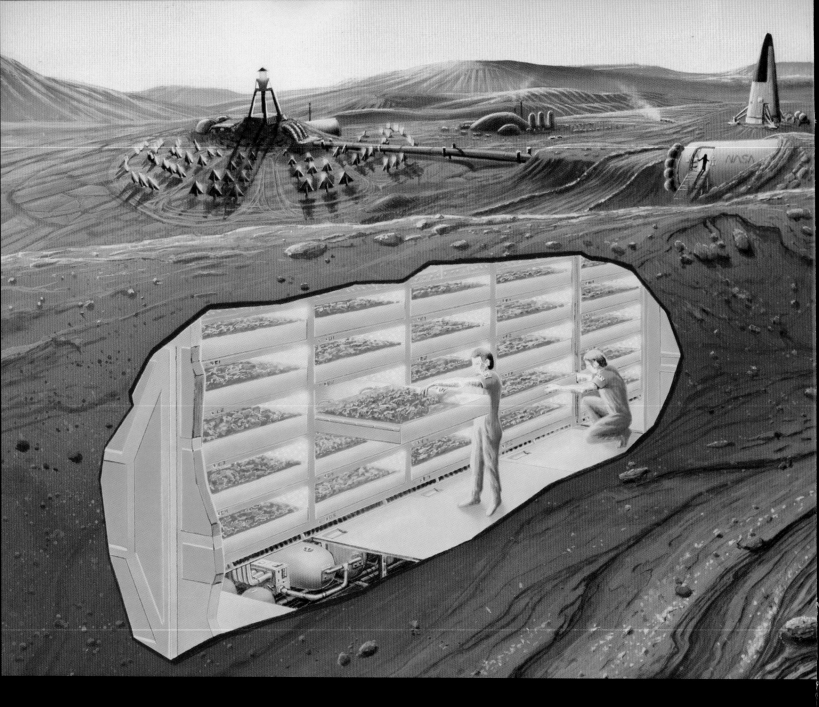

## Working the Land

The Curiosity rover's recent confirmation that the Martian soil contains large amounts of water even at equatorial latitudes paves the way for so-called "In-Situ Resource Utilization" (ISRU)—the manufacture of fuel for the return journey, alongside some other key supplies, on the surface of Mars.

Water is an ideal raw material for manufacturing liquid hydrogen and liquid oxygen, a highly combustible combination of propellants that generates powerful thrust when mixed together in a rocket engine. These same fuels could be utilized to generate power on a smaller scale. Properly processed, water itself could be used to replenish drinking supplies brought from Earth and, of course, it could also supply oxygen for breathing. It's little wonder that ISRU has become a hot topic among Mars enthusiasts. Some proposed missions call for a solar-powered fuel plant to be carried aboard the manned landing craft, while others involve landing an

automatic solar-powered fuel manufacturing plant first, to build up a fuel dump in advance of the astronauts' arrival. As a first step in turning the idea into a reality, NASA plans that their 2020 Mars rover mission will carry the Mars Oxygen ISRU Experiment (MOXIE), a prototype device for manufacturing oxygen on the Red Planet.

Harnessing Martian resources would make a longer stay on Mars, instead of a brief stopover sandwiched between two long and dangerous journeys, much more feasible, as any long-term mission would require agriculture to produce food and additional oxygen. Plants grown on Mars would rely on the same photosynthesis process they use on Earth, and would require carbon dioxide, nutrients, water, and sunlight.

Clearly, Mars has plentiful carbon dioxide and water, but terrestrial plants exposed on the planet's surface could not survive either the cold temperatures or the low atmospheric

A large-scale Martian power station like the one depicted here might be some way in the future, but it now seems likely that In-Situ Resource Utilization to generate electricity, water, oxygen, and even rocket fuel would play a key role in the manned exploration of Mars from the outset.

This artist's impression shows a future Mars outpost with habitation modules largely buried underground for temperature regulation and protection from dangerous radiation. Plants cultivated in hydroponic greenhouses would not only provide food for the crew, but also naturally convert carbon dioxide into oxygen. The semicircular structure in the background at upper left is a solar thermal power station, in which an array of mirrors called heliostats direct sunlight to the top of a tower containing a steam turbine.

pressure, so in practice some kind of greenhouse enclosure would be required. This would need to have an artificial atmosphere and soil enriched by nutrients brought from Earth. Even light presents a problem. The lower levels of sunlight on Mars, which are roughly half those received on Earth, would also be a major drawback for any transplanted crops. NASA scientists have been studying ways of using solar-powered LED lighting to provide only the specific wavelengths of light that photosynthesis requires.

## One-way Mission

As with the idea of ISRU fuel plants, sustainable agriculture on Mars presents a huge challenge, requiring extensive hardware. Many schemes for Mars missions involve sending unmanned vehicles and components ahead, ensuring they are safely operational on the surface before the manned spacecraft sets off. The alternative scenarios, in which problems develop with one of the mission components once the astronauts are already committed to their journey, or even stranded on the surface, are the stuff of science-fiction nightmares.

Some Mars enthusiasts, however, actively embrace the idea of a one-way trip with no guarantee of return. A "Mars to stay" mission, they argue, could be achieved with near-current technology at much lower costs, simply by avoiding the issues associated with putting a fully fueled return vehicle on Mars.

The Netherlands-based MarsOne organization has even got as far as recruiting astronaut candidates for a proposed mission from among the general public. Their ambitious plans aim to send a four-person crew to Mars by 2025, to be joined by further groups of "colonists" at regular intervals. Even if the ethical questions about sending people on a dangerous mission, ultimately to die on Mars, are set to one side, the MarsOne promoters seem at the very least to underestimate the demands of sustaining a long-term colony.

## Settlers on Mars

Despite all of the problems that will need to be resolved, it seems inevitable that, at some point in the future, intermittent manned missions will slowly evolve into a permanently manned Martian settlement.

There are many practical reasons for establishing a base on Mars, not least because it helps to address some of the problems of sending crews on such a long spaceflight. Once hardware such as fuel plants and greenhouses are up and running, maintained by personnel on the planet, there are huge benefits to reusing them on subsequent flights.

Even if astronauts do not remain beyond a single "tour of duty," which might involve spending two years on Mars, it seems likely that the first Martian landing site will become a *de facto* outpost with shifting personnel, similar perhaps to present-day scientific bases in Antarctica.

Longer-term survival in the hostile Martian conditions will almost certainly require such settlements to be underground, perhaps utilizing natural caves and hollow lava tubes that have recently been discovered in orbital photographs. These environments will not only provide more stable temperatures, but also shield settlers from the harmful effects of cosmic rays.

The need for reliable solar energy will probably restrict the first major Martian colonies to the equatorial plains, but such a permanent outpost will still open the way for a more thorough exploration of the entire planet. This will undoubtedly be a slow process as the surface of Mars covers an area only slightly smaller than that of all Earth's continents put together. Yet, as the colony grows, robot rovers, low-powered rockets for suborbital hops, and perhaps even piloted aircraft will make it easier for Martian explorers to venture farther afield.

Key scientific questions for these early colonists will revolve around the mysteries of the planet's history, particularly whether native Martian life ever existed, and whether it still does. But the Red Planet also represents a huge opportunity for life on Earth, allowing us to begin considering the long-term survival of humanity.

**ABOVE** This illustration of an astronaut abseiling down a steep Martian cliff formed the frontispiece to a 1997 NASA study of options for manned exploration of Mars. But while it captures the daring spirit of man's first adventure on the Red Planet, such exploits would be asking for trouble in the early days of a Martian outpost—astronauts would be weak and vulnerable from the long interplanetary voyage, with only limited medical equipment at hand and no hope of a quick route home.

**RIGHT** Another Mars outpost concept (from a 1989 report) depicts astronauts at work around a long-duration settlement including solid cylindrical sections, an inflatable dome, and unpressurized, Moon-buggy-style rovers. In the background, a drilling rig can be seen at work, while a recently released meteorological balloon floats up into the sky.

# Heading for the Stars

Our fragile little planet is an isolated ball of rock floating in the vastness of space, where life flourishes on and within a thin surface layer of soil, water, and air, existing in an atmosphere that is increasingly vulnerable to all manner of threats. From a cosmic perspective, we are frighteningly exposed to a wide range of dangers. Setting aside human politics and the weaponry we have developed that holds the potential to make our planet uninhabitable, there are natural disasters to consider that could just as easily devastate our world.

Life on Earth has suffered from mass-extinctions in the past due to climate change, volcanic activity, and impacts from space. We must expect that an event or events of an equally destructive nature will happen again in the future.

Learning how to survive on Mars and creating the technologies required to do so could be the best way for us to learn how to survive on our own planet as well as alien worlds.

Forging ahead with developing the science and engineering essential to further exploration of space will become key to our survival. No less a figure than cosmologist Stephen Hawking has pointed out that, if we want to ensure the long-term future of our species, a colony on Mars is the obvious first step.

It will be a long and difficult road, with many setbacks along the way, but one day, perhaps a century from now, there really will be Martians on the Red Planet.

ABOVE A sequence of images depicts a far-future Mars in the various stages of terraforming that could ultimately produce a more hospitable, Earthlike planet.

## A Blue Planet?

One of the most obvious things that separates humans from most other species on Earth is the way we reshape the environment around ourselves, altering the landscape, resource supplies, and even other species to better meet our own requirements. Once a permanent human settlement is established on Mars, there seems little doubt that we will begin to do the same thing there.

Present-day Mars might be the most hospitable of the other worlds in the solar system, but it is still unremittingly hostile to what we might consider normal life. The temperatures rarely rise above freezing; water stays locked away in the soil and ice caps; the atmosphere is both thin and toxic to animal life; and dangerous radiations from space rain down on the surface.

While early settlers may be content to live their lives in pressurized habitation modules or converted caves and sinkholes, only ever venturing out in the protection of a

heavy spacesuit, at some point their descendants may want to stretch their legs. Should venturing outdoors without adequate protection mean sentencing yourself to death, or is there a potential alternative?

Fortunately, scientists and science fiction writers have been thinking about this challenge for almost a century now, developing an entire field of theoretical planetary engineering known as terraforming. The astronomer and science communicator Carl Sagan was the first to pay serious attention to the question of terraforming Mars in a scientific paper as long ago as 1973.

In the case of the Red Planet, we now know that many of the raw ingredients for a more Earthlike world are locked away in the soil and ice caps. The first step, therefore, would be to generate a thicker atmosphere that could warm the planet slightly, causing carbon dioxide to sublimate out of the permanent ice caps, thus further thickening the air. Once the envelope of gases reached critical density and about one-third of Earth's atmospheric

pressure, humans would be able to walk on the surface without protective spacesuits, even though the atmosphere would remain toxic.

## A Breath of Fresh Air

A stroll outdoors would still require breathing apparatus but, over time, the heat-retaining properties of the carbon-dioxide-rich atmosphere would create a greenhouse effect to further warm the planet, until eventually water ice began to melt and water ran on the surface.

Meanwhile, hardy plants, perhaps genetically engineered to cope with the reduced sunlight and encouraged to take root in the Martian soil, would start to naturally convert carbon dioxide into the free oxygen necessary for animal and human life.

More oxygen could be manufactured artificially by processing water or the perchlorate salts that appear to be widespread in the Martian soil.

Needless to say, this brief summary is grossly oversimplified and the challenges associated with such a scheme are immense. One of the biggest is simply how to hold on to the enriched atmosphere. Warmer gases will tend to billow out from the planet, escaping its weak gravity, or be carried off on the solar winds that batter Mars thanks to its feeble magnetosphere.

Of course, the question of how to achieve the initial enrichment that kickstarts the process is also hugely problematic. Suggestions range from bombarding the surface with asteroids to introducing CFCs, the highly effective artificial greenhouse gases that mankind has spent recent decades trying to eradicate.

The costs of terraforming Mars are unfathomable, and the timescales involved are measured in millennia rather than years, but perhaps one day, in the far distant future, the inhabitants of Earth will look up into the sky and see not a Red Planet, but a blue one.

# INDEX

## Picture Credits

Key: a-above; b-below.

10–11, 14–15, 40–41, 46–47, 48, 58–59, 61, 62, 63, 73, 76–77, 85a, 86b, 130–31 ESA; 6 Captmondo/Creative Commons Sharealike; 8 Jean-Pol Grandmont/Creative Commons; 126a Brocken Inaglory/Creative Commons Sharealike; 129 Nesnad/Creative Commons Sharealike

All other images supplied courtesy of NASA.